HOME BUILDERS' Safety Program

NATIONAL ASSOCIATION OF HOME BUILDERS

HOME BUILDERS' Safety Program

Home Builders' Safety Program

	NAHB Labor, Safety & Health Services
Robert Matuga	Director
Kevin Cannon	Safety Specialist

BuilderBooks, a Service of the National Association of Home Builders

Christine B. Charlip	Publisher
Courtenay S. Brown	Book Editor
Torrie Singletary	Production Editor
Circle Graphics	Cover Design & Composition
Victor Graphics, Inc.	Printing
Gerald M. Howard	NAHB Executive Vice President and CEO
Mark Pursell	NAHB Senior Staff Vice President, Marketing & Sales Group
Lakisha Campbell	NAHB Staff Vice President, Publications & Affinity Programs

Disclaimer

This publication provides accurate information on the subject matter covered. The publisher is selling it with the understanding that the publisher is not providing legal, accounting, or other professional service. If you need legal advice or other expert assistance, obtain the services of a qualified professional experienced in the subject matter involved. Reference herein to any specific commercial products, process, or service by trade name, trademark, manufacturer, or otherwise does not necessarily constitute or imply its endorsement, recommendation, or favored status by the National Association of Home Builders. The views and opinions of the author expressed in this publication do not necessarily state or reflect those of the National Association of Home Builders, and they shall not be used to advertise or endorse a product.

©2007 by BuilderBooks. All rights reserved. No part of this book may be reproduced or utilized in any form or by any means, electronic or mechanical, including photocopying and recording or by any information storage and retrieval system without permission in writing from the publisher.

Printed in the United States of America

11 10 09 08 07   1 2 3 4 5

ISBN 978-0-86718-630-7

Cataloging-in-Publication Information

Library of Congress Cataloging-in-Publication Data

Home Builders' Safety Program / NAHB Labor, Safety and Health Services Department.
    p. cm.
  ISBN-13: 978-0-86718-630-7 (alk. paper)
  ISBN-10: 0-86718-630-5 (alk. paper)
 1. House construction—Safety measures. 2. Building—Safety measures.
I. National Association of Home Builders (U.S.). Labor, Safety, and Health Services
Dept. II. Title.

 TH443.H66 2006
 690'.80288—dc22

2006051651

For further information, please contact:

BuilderBooks.com — A Service of NAHB
BOOKS THAT BUILD YOUR BUSINESS
1201 15th Street, NW
Washington, DC 20005-2800
800-223-2665
Visit us online at www.BuilderBooks.com.

Contents

Acknowledgments IX

Disclaimer XI

How to Use This Book XIII
    What You Will Find on the CD XIII

CHAPTER

1. Overview of Jobsite Safety 1

    Major Safety Motivators 1
    Safety Starts at the Top 2
    Insurance and Workers' Compensation 2
    Marketing Your Safety Program 3

2. Understanding the Occupational Safety and Health Act 5

    Construction Standards and General Industry Standards 5
    Current OSHA Inspection Policy 6
    What to Do When the OSHA Inspector Knocks at Your Door 8
    OSHA Inspection Action Plan 8
    The Aftermath of an Inspection 11
    OSHA Closing Conference Guide 11
    Multiemployer Citation Policy 14

3. Model Safety Program for Builders 17

    State Goals in Writing 18
    Write an Action Plan 18
    Establish a Budget 18
    Designate a "Safety Champion" 18
    Define Areas of Responsibility 19
    Develop and Implement Jobsite Safe Work Practices 21
    Establish Accountability Procedures 21
    Develop and Deliver a Safety Training Program 21
    Develop and Implement Enforcement Procedures 23
    Conduct Regular Jobsite Inspections and Hazard Analysis 23
    Develop Recordkeeping Procedures 25

Establish Accident Reporting and Investigation
    Requirements   26
Develop and Maintain a Trade Contractor Safety Compliance Policy   27
Evaluate the Effectiveness of Your Safety Program   28
Encourage Feedback and Reward Excellence   28

4. OSHA Regulations and Safe Work Practices   29

Personal Protective Equipment   29
Fire Protection and Prevention   31
Tools   34
Welding and Cutting   36
Electrical Safety   39
Scaffolds   41
Fall Protection   49
Cranes   61
Motor Vehicles   63
Excavation and Trenching   64
Stairways and Ladders   66
The Hazard Communication Standard (HazCom)   68

5. Safety and Health Training   73

OSHA Regulations Requiring Training for Compliance   73
Stated Requirements   73
Implied Training Requirements   75
Adopted Training Requirements   76
New Hire Orientation   76
Recording Training History   76
Ongoing Training and Awareness   77

APPENDIXES

A. Additional Sources of Information   81

B. OSHA Regional Offices and States with Approved Programs   83

C. Appendix C: Forms and Checklists   87

Employee Safety Violation Reprimand Form   88
Accident Investigation Report Form   89

CONTENTS

VII

    Construction Safety Checklist   91
    New Hire/Trade Contractor Safety Orientation Checklist   94
    Safety Program Checklist   95

D. Model Safety Program   97

E. Model Written Hazard Communication Program   123

F. Model Hazard Communication Employee Training Program   125

Acknowledgments

NAHB Labor, Safety, & Health Services gives special acknowledgments to the following individuals who assisted in the development of the *Home Builders' Safety Program*: Matt Murphy, Safety Environmental Engineering; Vern Pottenger, Chairman, and Andy Anderson, Vice-Chairman, NAHB's Construction Safety & Health Committee; Kevin Cannon, Safety Specialist, NAHB Labor, Safety, & Health Services; Christine Charlip, Publisher, and Courtenay Brown, Editor, NAHB BuilderBooks. *The Home Builders' Safety Program* was prepared under the general direction of NAHB's Director of Labor, Safety, & Health Services, Rob Matuga.

We thank all those who contributed to the earlier edition of this publication.

Disclaimer

The model safety program presented in this book and on the accompanying CD should be used only as a guide for developing your company's safety program. Circumstances can vary widely, so modify the program to reflect your company's actual operations.

The sample language and forms are provided for informational and educational purposes only. Forms may not contain every provision that is necessary for a particular construction project, and they may also contain some provisions that do not apply. The forms may not necessarily be compatible with the laws of each state or locality. Applicable law differs widely among the states, and in certain cases, local municipal law may apply. If conflict arises between these forms and requirements of locally approved codes or locally approved or accepted guidelines, as a matter of law, the code requirements or performance criteria may take precedence over these forms. Consult your attorney concerning the valid use of these documents in your jurisdiction and concerning the appropriate language and format of form instruments.

How to Use This Book

This building industry model safety program is designed for small to mid-sized residential construction companies to use as a practical guide for developing an effective in-house safety program.

This building industry model safety program is designed for small to mid-sized residential construction companies to use as a practical guide for developing an effective in-house safety program. This book will guide you through the process of developing and implementing a safety program.

- Chapter 1 provides an overview of jobsite safety. You will learn why a safety program is critical to your company's overall success.

- Chapter 2 explains the Occupational Safety & Health Act. You will learn what your company needs to do to be in compliance. This chapter will also help you create guidelines for handling an Occupational Safety & Health Administration (OSHA) inspection. The *Multiemployer Citation Policy* is also discussed in this chapter.

- Chapter 3 explains the components of the safety program. This program does more than just cover compliance with OSHA standards. It is the next generation of safety programs: a total loss-control program. This program is designed to be adaptable—not all of its sections are necessarily applicable to every builder. A checklist to assist you in developing a safety program for your company is included in Appendix C and on the CD that accompanies this book.

- Chapter 4 provides comprehensive safe work practices and OSHA requirements for residential contractors. Your safety program should include the safe work practices for the conditions on your jobsites. These practices are included on the CD that accompanies this book for ease in incorporating into your safety program.

- Chapter 5 discusses the need for safety training. It provides an outline for new hire orientation and stresses the importance of ongoing training.

WHAT YOU WILL FIND ON THE CD

The CD that accompanies this book contains the following materials:

- Model Safety Program
- Safety Program Checklist
- Model Written Hazard Communication Program

- Model Hazard Communication Employee Training Program
- Employee Safety Violation Reprimand Form
- Accident Investigation Report Form
- Construction Safety Checklist
- New Hire/Trade Contractor Safety Orientation Checklist

The files can be easily customized to meet your business needs. To access the files

1. Place the CD in your CD drive.
2. Launch Microsoft® Windows Explorer.
3. You will see three subdirectories:
   - Word Files: These files can be edited using Microsoft® Word.
   - Text Files: These files allow you to customize the forms to suit your specific needs.
   - PDF Files: These files are pre-designed forms that you can immediately print and use.
4. Double-click the icon that represents your CD drive.
5. Select Copy and copy the files to your hard drive.

1

Overview of Jobsite Safety

This chapter provides an overview of jobsite safety. You will learn why a safety program is critical to your company's overall success.

It is an indisputable fact: Construction work can be dangerous. Therefore, employers and employees must take all available precautions to safeguard against accidents and injuries that can cost lives and drive companies into bankruptcy.

Many builders and contractors are either unaware of the dangers or simply ignore them in favor of believing that the next accident "will happen to the other guy, not me." This kind of faulty thinking puts lives and companies at risk.

Most construction accidents are not caused by building collapses or crane malfunctions on large jobsites. Rather, it's the trench collapse, the improperly erected scaffolding, or an employee's failure to wear a safety harness or operate equipment correctly. These types of problems plague all builders and contractors—large and small, residential and commercial.

Builders need to appreciate the value of a good safety program. A good safety program saves lives and protects your bottom line by reducing

- workers' compensation costs
- equipment losses
- time spent on filing accident claims and reports
- injured worker replacements
- periods of lowered or decreased productivity

MAJOR SAFETY MOTIVATORS

Unfortunately, many construction companies take a *reactive* approach to safety—their safety program takes effect only after an accident has already occurred or as they prepare for a visit from OSHA. Studies show that the best safety programs are based on a *proactive* approach. There are three major motivator types for safety:

- **Humanitarian.** We don't want to see our workers hurt.
- **Economical.** We don't want to incur large fines, doctor bills, increased insurance rates, or productivity losses.
- **Legal.** We don't want to have to deal with lawsuits brought by workers who were injured on our jobsite.

No matter which motivator prompts you to action, it will take a long-term commitment to establish a good safety plan for your

Common Causes of Jobsite Accidents

- lack of safety management
- lack of safety enforcement
- lack of continuing education measures
- unsafe work practices
- tight deadlines
- lack of responsibility
- lack of proactive jobsite safety

company. Safety programs are an important and sound investment in your business. The more effort you invest in developing a sound program, the greater your future returns.

SAFETY STARTS AT THE TOP

If management does not make safety a priority, then neither will the rank and file. Simply telling people to practice job safety techniques is not enough. You must educate your employees on how to *be* safe and how to *think* safe. Preaching safety without teaching it is like telling your workforce to build the project without providing the necessary plans or directions.

Safety must be an integral part of your workplace culture. Otherwise, you are condoning dangerous working conditions.

Actual Cost of an Accident: Direct and Indirect

Direct costs
- damage to property
- cost of medical bills
- time off
- citations

Indirect costs
- lost time on the project
- insurance premium increases
- loss of a person who knows the project scope
- training for replacement workers
- possible legal fees
- long-term disability
- lost profits

Often workers criticize those who practice safety measures like wearing a hardhat on the jobsite. If management mandates that all workers must wear hardhats on the jobsite, then no one can be singled out and ridiculed. Job safety measures must be ingrained into your workforce, so that the workplace culture promotes safety.

Shorter deadlines are spawning an increase in accidents. Workers are doing everything possible to deliver the product on time, including cutting corners. For example, a crew that is rushing to set trusses for a roof may not install the required temporary bracing, allowing for the possibility of a roof collapse resulting in serious injuries or fatalities.

Your company mandate must be that everyone on the project is responsible for safety. Important safety measures, such as installing guardrails, are often left undone because the prevailing attitude is that safety is someone else's responsibility. However the fact is that each employee is key to the success of your safety program. Therefore, everyone must be able to identify and correct potential safety issues.

Prevention is the key to jobsite safety. All too often, when accidents occur workers will scramble to get help for the injured person and then try to make the site appear as if it had been safe. A safe jobsite makes accidents less likely to happen at all.

INSURANCE AND WORKERS' COMPENSATION

How do you calculate your workers' compensation insurance rate? Insurance companies look at the number of accidents that have occurred in the previous three years to calculate your experience modifier rate

(EMR). This rate sets your premiums. The fewer the number of accidents, the lower the premiums you pay.

If you can reduce the number of accidents on your jobsite for the next three years, you can decrease your worker compensation rates (Table 1).

MARKETING YOUR SAFETY PROGRAM

Builders can market their safety success not only to future home owners but to their insurance agents (when shopping for lower rates) and prospective employees. A company that has a strong reputation for safety tends to attract quality employees.

Trade contractors can market their safety programs to builders. By training their employees and providing the correct materials for jobs, there is less likelihood of accidents and less need for supervision of their overall safety. This saves the contractor and the builder time and lessens the possibility of citations and lawsuits.

A work culture that promotes safety helps you attract the best people to your firm. Prospective employees take notice of where a firm is on the safety continuum. At one end is the well-rounded firm that protects its employees by implementing and enforcing safety programs. These firms care about their brand and their employees' well-being and produce a great product, so they tend to stay in business longer. At the other end, is the "get-it-done and move-on" type of firm that rushes through the job, is careless about safety, and is usually besieged by major mistakes and accidents. Who wants to work for this type of firm when they can work for a safe, quality builder?

Table 1. Potential Savings Based on Experience Modifier Rate

	Company A	Company B
Payroll	$100,000	$100,000
Manual rate	× 0.15	× 0.15
EMR	$15,000 × **1.68**	$15,000 × **.65**
Workers' comp rate	$25,200	$9,750
Insurance Savings for Company B		**$15,450**

2

Understanding the Occupational Safety and Health Act

This chapter explains the Occupational Safety and Health Act. You will learn what your company needs to do to be in compliance. This chapter will help you create guidelines for handling an Occupational Safety and Health Administration (OSHA) inspection. The Multiemployer Citation Policy is also discussed.

The Occupational Safety and Health (OSH) Act of 1970 is among the most comprehensive legislation ever enacted regarding the regulation of hazards in the workplace. Since the Act went into effect on April 28, 1971, more than 50% of all scheduled inspections conducted by both federal and state Occupational Safety and Health Administration (OSHA) agencies have been targeted at the construction industry.

In the Act, the *General Duty Clause* basically states that all workplaces shall be maintained free from recognized hazards. Once hazards are recognized, the employer must take the proper steps to eliminate those hazards.

By choosing a career in the construction industry, you chose to abide by OSHA rules. The OSHA 1926 Construction Standards are a minimum set of rules to increase worker protection. All construction industry companies have a duty to comply with these standards. No company is exempt from the rules set by OSHA.

Several states have their own OSHA programs. State programs must be at least as effective as the federal standards, meaning that states must enforce the federal standards, but have the discretion to raise the safety bar in their programs. State programs cannot accept anything less than the federal standards. Appendix B includes a list of state programs and contact information.

CONSTRUCTION STANDARDS AND GENERAL INDUSTRY STANDARDS

There are two sets of standards that builders must follow in order to be considered in compliance with OSHA: the 1926 Standards for Construction and the 1910 Standards for General Industry. The 1926 Standards are the primary construction standards for safety compliance in the building industry. The 1926 Standards reference the 1910 Standards as well as consensus standards that pertain to employee safety and protection from the American National Standards Institute (ANSI), the National Institute for Occupational Safety and Health (NIOSH), and the National Fire Protection Association (NFPA), to name a few. For example, Respirator Protection (1926.103) states that the protection requirements are the same as 1910.134; therefore, the employer must comply with 1910.134. Standards pertaining to foot protection are found in 1926.96, which states that foot protection must meet

Past OSHA Enforcement Actions

Historically, federal and state OSHA inspections in the construction industry have focused on the following issues:

- projects involving a large number of contractors and trade contractors, such as a subdivision development
- projects involving trenching and excavations, tunnels and shafts, asbestos removal, and other high hazard operations
- follow-up to employee complaints of alleged company violations, especially concerning "imminent danger" working conditions
- fatal accidents and catastrophe investigations

In short, if you are working on large or dangerous projects, or projects connected to complaints or investigations, your chances of being inspected by OSHA are greatly increased.

ANSI Z41.1-1967. Referenced standards must be followed to reach the minimum standards set forth by OSHA.

The Act expressly denies an employer advance warning of inspections, thus allowing for surprise visits by OSHA representatives. Therefore, it is imperative that the builder knows and understands his or her rights in advance. For example, you have the right to request a warrant before admitting an inspector onto a jobsite. Few builders take advantage of this option out of fear of retaliation. In this chapter, you will learn

- how to respond to an OSHA inspection
- what OSHA inspectors look for during an inspection
- your rights during and after an inspection
- your options and recourse in response to a citation
- your responsibility for trade contractor violations

Builders and contractors are strongly urged to obtain a copy of the most recent edition of *OSHA's Construction Safety and Health Standards* (29 CFR 1926). It is available at BuilderBooks.com, which offers a variety of safety and regulation compliance materials (www.BuilderBooks.com or 1-800-223-2665).

CURRENT OSHA INSPECTION POLICY

OSHA has instituted a "focused" inspection policy for the construction industry (residential and commercial) that was a drastic change from its previous inspection process. Since then, OSHA has focused on the four major hazards that result in the most serious construction injuries for all programmed and referral inspections of the construction industry. The new policy is in effect for federal OSHA program states. The "big four" are the leading hazards that could cause employee injury or death:

1. falls from elevations
2. electrical shock
3. caught in/between
4. struck by

The OSHA inspection process is a two-step process. First, the Compliance Safety and Health Officer (CSHO), also known as the inspector, determines whether 1) the builder has an effective written safety and health program that meets OSHA guidelines, and 2) a responsible, capable individual assigned to implement the program. Second, the inspector determines whether any of the big four hazards exist on the jobsite.

The big four hazards account for most of the fatalities and serious injuries in the building industry. For years, the construction industry had argued that small business owners have wasted time, energy, and resources to comply with OSHA standards, such as the Hazard Communication Standard that had little to do with fatalities on the jobsite. Therefore, OSHA's new, focused inspection policy is one of the most significant advances in safety for the home building industry in recent history.

What does this focused inspection policy mean when OSHA shows up for an inspection? It's hard to guarantee exactly how the inspection process will evolve, but it will probably unfold as follows:

> Bob, a small home builder, was a conscientious employer and had always cared about jobsite safety. Bob instituted a full safety and health program for his employees after attending an educational seminar on safety. In addition, Bob required all his trade contractors to institute an equally effective safety program. The contractors followed all of the OSHA regulations; for example, the roofer installed and used anchor points and lanyards, and the excavator protected against cave-ins. Bob ensured that the employees were protected from exposure to live electrical parts, and that all vehicles were routed to prevent hitting employees. In addition, Bob ensured that all employees used ground fault protection. Even though Bob had an effective safety program, he was always afraid that one of his trade contractors or employees would bring a new product on site without all of the required material safety data sheets (MSDS).
>
> On a particularly busy morning, an OSHA inspector showed up at the jobsite. Fortunately, the site supervisor was present and called Bob to meet them at the site. After all the introductions, Bob retrieved his safety program from his job box and explained that he was the safety program administrator. After the inspector reviewed the program, he conducted a jobsite "walk around." The inspector witnessed how the written program was being effectively implemented on the site and the steps to lessen the big four hazard risks were in compliance with OSHA standards. A few hazards were noted but not cited because Bob fixed them on the spot. The inspector closed the inspection, waved goodbye, and no penalties were assessed. Bob's hard work had paid off.
>
> Later, Bob heard about another builder in the area who spent thousands of dollars to hire a safety consultant to write a safety program. The program looked great and contained all of the necessary requirements, but the builder never implemented it. When OSHA inspected that builder, the inspection was expanded to the fullest possible extent because an effective written safety program was not in place.

According to OSHA, the goal of its new policy is to invest as few resources as possible on those employers who develop safe worksites.

WHAT TO DO WHEN THE OSHA INSPECTOR KNOCKS AT YOUR DOOR

You should understand your company's rights under the OSH Act as part of your company's program to prepare for a federal or state OSHA inspection. Your employees should be trained on how to handle an OSHA inspection. You should have written processes for getting proper personnel on site and how to respond during an inspection. For example, if your president wants to walk the jobsite with the inspector and it is documented in a written plan, the inspector will often allow a reasonable amount of time for that person to arrive on site.

The following action plan is a practical guide to help you, your superintendents, or your designated representative deal with OSHA inspectors. It offers brief interpretations of existing law intended to explain your company's rights during an inspection. As new regulations and interpretations are issued by the Department of Labor, some of these guidelines may change. Consider incorporating this plan in your company safety program and distributing it to your management team so that they understand the inspection process and your company's due process rights.

OSHA INSPECTION ACTION PLAN

1. **Be polite, respectful, and cooperative.** Hostile attitudes and attempts to delay or interfere with the investigation will only result in your company losing precious rights during the inspection and receiving maximum penalties and fines for any violations. The atmosphere of the investigation should be a cooperative one.

2. **Verify the inspector's credentials.** You are not required to allow the inspector onto the jobsite unless he or she presents proper identification. The OSH Act specifically provides that "upon presenting appropriate credentials to the . . . agent in charge," the inspector should be allowed to enter the workplace without delay. This means that the company's highest-ranking official is entitled to see the inspector's identification papers to verify authenticity. If the site superintendent is not present, then a designated representative must receive the inspector. It is reasonable to ask the inspector to wait a few minutes to allow your company's highest-ranking official to arrive. However, do not use this right as a means of delaying the inspector's entry. An inspector who presents proper credentials must never be denied entry onto the jobsite.

3. **Participate in a brief preinvestigation conference.** Before the inspection begins, the inspector will explain the nature and purpose of the inspection and identify the records he or she wishes to review and the employees he or she wishes to question. Although this summary will not preclude any additional investigations, it will provide a guideline to help you assist the inspector in conducting an efficient,

orderly, and fair inspection. You can request time to advise the company owner of the inspection as long as it does not delay or interfere with the inspection.

4. **Get the inspector's business card and copies of complaint(s).** It is important to record all relevant information concerning the inspection in case your company decides to contest a citation. If the investigation is the result of a written complaint, you should get a copy of the complaint. Request a business card from everyone who was present during the inspection. Under the law, the inspector cannot release the names of the complainant. However, it may be important for you to know whether outside interests are attempting to use the safety inspector to disrupt the job. Try to ascertain if the inspection is routine by asking if the complaint was filed by one of your employees, an employee of another contractor, or an outside party. Avoid the appearance of guessing the identity of the complainant. If the inspector refuses to tell you, then you should drop the subject.

5. **Make the inspector aware that you know reasonableness is your right.** The OSH Act guarantees that the inspection will be reasonable, orderly, and fair. This means the inspections must occur at a reasonable time, within reasonable limits, and in a reasonable manner, of such places of employment and all pertinent conditions, structures, machines, apparatus, devices, equipment, and materials. The test of reasonableness should be if, after the preliminary inspection, the requests by the inspector for further examination or questioning are grounded in a reasonable belief that further examination or questioning would reveal an unsafe or unhealthy condition. If you believe that a request is unreasonable, explain your concerns to the inspector. If the inspector insists, you can comply or request time to consult the company owner. If you strongly believe that the request is unreasonable and unnecessary, consult the company owner before proceeding. There may be other areas the inspector can inspect while a decision is being made on the area in question.

Under the OSH Act, the inspector may interview employees privately and may also examine any machinery or equipment in the workplace. The inspector also has the right to take pictures and samples and to use other reasonable investigative techniques.

6. **Avoid disruptions of work in progress.** As a part of the requirement that an inspection be conducted in a reasonable manner, OSHA guidelines direct the inspector to conduct the investigation so as to avoid any undue and unnecessary disruption of the normal operations of the employer. It is your duty to inform the inspector of the day's construction schedule and to assist him or her in conducting the investigation so as not to unduly interfere with the work.

7. **Accompany the inspector.** You, or your authorized designee, have the right to accompany the inspector during the physical inspection. However, OSHA regulations expressly provide the inspector with the authority to deny the right of accompaniment to any person whose conduct interferes with a fair and orderly inspection. In the event that an authorized employee representative is not available to accompany the inspector, the inspector will consult with a reasonable number of employees concerning matters of health and safety in the workplace during the inspection. Your company is protected under this provision from outside interference in the inspection by individuals who claim to represent the interest of the employees but who have not been duly designated as the employees' representative.

8. **Take notes and pictures.** As you accompany the inspector during the inspection, it is imperative to take notes and pictures from the same position as the inspector, if possible. You, or your representative, should prepare a written report incorporating your notes and photos and any relevant comments by the inspector or other information acquired during the pre-inspection and postinspection conferences and the actual inspection.

9. **Participate in a postinspection conference.** On completion of the inspection, the inspector must confer with the site superintendent to advise him or her of any apparent safety or health violations disclosed by the investigation. Therefore, it would be advantageous to have a person authorized to make decisions present at this conference. On jobsites where employees are represented by an authorized representative, the closing conference can be held jointly with the employer and employee representative; however, this decision is at the discretion of the inspector.

10. **The inspector may declare imminent danger.** If the inspector concludes that conditions or practices exist that could reasonably be expected to cause death or serious harm before the danger can be eliminated, he or she may ask you to abate the danger immediately. If the danger can be immediately abated without incurring great expense or halting work on the jobsite, then you should promptly correct the problem. The inspector cannot close the jobsite without a court order, so you have time to consult the company owner. If the owner decides not to abate the danger without a court order, the inspector must inform the affected employees of the danger and advise the company owner and the employees that he or she is recommending a civil action to restrain or remove such conditions and then leave the site. If the danger proves to be a violation of the OSH Act, or if an employee is injured or killed before a court order can remove the danger, your company officials will have opened themselves to potential criminal penalties.

THE AFTERMATH OF AN INSPECTION

Assume the inspector has found conditions that may be in violation of the OSH Act. After the information regarding these conditions is examined by the inspector's supervisor, he or she may agree that violations exist. If so, the inspector may issue citations, explain in detail the exact nature of these violations, and set forth any associated penalties.

For each apparent violation found during the inspection, the inspector should have discussed or plan to discuss the following issues with the company's representative:

- the nature of the violation
- possible abatement measures needed to correct the violation
- possible abatement dates you may be required to meet

Note that certain hazards may also have been found during the inspection that may require further examination by another compliance officer. For example, a suspected occupational health hazard may require evaluation by an industrial hygienist.

The following closing conference guide should also be given to management personnel involved with the inspection.

OSHA CLOSING CONFERENCE GUIDE

During the closing conference, your company will be informed of all hazards that may be referred to another inspector for examination at a later date. Any such examination is considered a continuation of the original inspection.

If your company receives a citation as a result of an inspection, consider the following information on your and OSHA's rights and obligations:

1. **Read and understand the citation and pamphlet.** OSHA will provide a copy of the *Employer Rights and Responsibilities Following an OSHA Inspection* pamphlet with the citation. (The pamphlet is also available online at www.osha.gov/publications/osha3000.html.) You should read the citation and the pamphlet carefully. If you have any questions, contact the OSHA area office at the address listed on the citation.

2. **The citation must be posted.** The actual citation, or a copy of it, must be posted at or near the place each violation occurred to inform employees about the hazards that they may be exposed to. The citation must remain posted for three working days or until the violation is corrected, whichever is longer. Weekends and federal holidays are not counted as working days. Your company must comply with these posting requirements even if you contest the citation.

3. **Comply with the citation and notification.** Your company must comply with the citation and notification of penalty unless you follow the appropriate contest procedure. If your company agrees to the citation and penalty, you must correct the condition by the date in the citation and pay the applicable penalty.

4. **You may contest any portion of the citation and notification of penalty.** You have 15 working days after receipt of a citation and notification of penalty to submit a written *notice of contest* to the OSHA area director. The notice must clearly state what is being contested—the citation, the penalty, the abatement date, or any combination. The pamphlet accompanying any citation offers additional details.

5. **The contest process involves a public hearing.** If you properly file your notice of contest, the OSHA area director will forward your case to the Occupational Safety and Health Review Commission (OSHRC). The commission then assigns the case to an administrative law judge for a public hearing. The administrative law judge may choose to uphold, modify, or eliminate any item of the citation or penalty.

6. **Your company may have an informal conference.** You may request an informal conference with the OSHA area director to

   - obtain a more complete understanding of the specific standards that apply
   - discuss ways to correct the violations
   - discuss questions concerning proposed penalties
   - discuss problems concerning employee safety practices
   - obtain answers to any other related questions you may have

   If a citation is issued, an informal conference or a request for an informal conference will not extend the 15-day period within which you must either pay penalties or elect to contest.

   If you have a valid reason to obtain a longer abatement date, you can discuss this with the area director in an informal conference. He or she may issue an amended citation changing an abatement date prior to the expiration of the 15-day period without your filing a contest. If your company is only contesting the penalty, you must still correct all violations by the dates indicated on the citation.

7. **You must pay penalties promptly.** Penalties must be paid within 15 working days after you receive the citation and notification of penalty. However, if you file a notice of contest, you will not need to pay for those items contested until a final decision is rendered.

8. **Advise the area director of any corrective action you take.** For violations you do not contest, you should promptly notify the OSHA area director in writing that you have corrected the cited conditions by the abatement date. Your notification should explain the specific action that was taken with regard to each violation and the approximate date the corrective action was completed. When the citation permits an extended time for abatement, you must ensure that employees are adequately protected during this time. For example, the citation may require the immediate use of personal protec-

tive equipment by employees while engineering controls are being installed. Your company also should send periodic progress reports on its actions to correct these violations.

9. **You may choose to seek modification of the abatement date.** Abatement dates are established on the basis of the information available at the time the citation is issued. When uncontrollable events or other circumstances prevent your meeting an abatement date and the 15-day period has expired, you may submit a petition for modification of abatement. Further information is included in the *Employer Rights and Responsibilities Following an OSHA Inspection* pamphlet.

10. **A follow-up inspection may be conducted.** If your company receives a citation, a follow-up inspection may be conducted to verify that you have

    - posted the citation as required
    - corrected the violations as required in the citation
    - adequately protected employees during the abatement period

    You also have an ongoing responsibility to comply with the OSH Act. Any new violations discovered during a follow-up inspection will be cited.

11. **OSHA will penalize your company for failing to correct a violation by the abatement date.** If you do not correct the uncontested violations noted in the citation, OSHA will penalize your company. Therefore, it is important to begin abatement efforts immediately.

12. **Do not provide false information.** Providing false information regarding your abatement efforts on cited conditions or in records required to be maintained is illegal and punishable under the Act.

13. **Employees may challenge abatement dates.** Employees or their authorized representative may contest any or all of the abatement dates set for violations if they believe them to be unreasonable.

14. **It is unlawful to discriminate against employees.** The Act prohibits employers from discharging or discriminating against an employee who has exercised any right under this law, including the right to make safety or health complaints or to request an OSHA inspection. Complaints from employees who believe they have been discriminated against will be investigated by OSHA. If the investigation discloses a probable violation of employee rights, court action will follow.

15. **Your company may seek a temporary variance from a standard.** The Act permits your company to apply to OSHA for a temporary variance from a newly issued standard if you are unable to comply by its effective date because of the unavailability of materials, equipment, or technical personnel. You also may apply for a permanent variance from a standard if you can prove that your facilities or methods of operation are at least as safe and healthful as that

required by the OSHA standard. Contact your OSHA area director for more information.

MULTIEMPLOYER CITATION POLICY

The relationship between builders and trade contractors tends to fluctuate during a construction project, and nothing places more strain on the relationship than an OSHA jobsite inspection. An increasing number of builders are finding out that when an OSHA inspection takes place on their jobsite, they can be held responsible for the violations of their trade contractors.

Shortly after the OSH Act went into effect in the early 1970s, OSHA adopted a policy to address the unique situations of the multiemployer worksite. The Multiemployer Citation Policy allows OSHA to cite more that one employer for a hazardous condition that violates an OSHA standard, where multiple employers are at work on a jobsite. The policy, referred to as Compliance Directive CPL 02-00-124– CPL 2-0.124, became effective on December 10, 1999. It clarifies OSHA's Multiemployer Citation Policy and suspends Chapter III. C. 6. of OSHA's Field Inspection Reference Manual. This policy states:

> On multiemployer worksites, more than one employer may be citable for a hazardous condition that violates an OSHA standard. A two-step process must be followed in determining whether more than one employer is to be cited.

Step One. The first step is to determine whether the employer is a creating employer (the employer that caused a hazardous condition), exposing employer (an employer whose own employees are exposed to the hazard), correcting employer (an employer who is responsible for correcting a hazard), or controlling employer (an employer who has general supervisory authority over the worksite, including the power to correct safety and health violations itself or require others to correct them). Once the role of the employer has been established, OSHA will go to step two to determine if a citation is appropriate.

Step Two. If the employer falls into one of the categories of creating, exposing, correcting, or controlling employer, it has obligations with respect to OSHA requirements. OSHA will then determine if the employer's actions were sufficient to meet those obligations. OSHA does note that the extent of the measures that a controlling employer must take to satisfy its duty to exercise reasonable care to prevent and detect violations *is less than* what is required of an employer with respect to protecting *its own* employees.

OSHA evaluates whether or not an employer's actions were sufficient to meet obligations with respect to OSHA requirements by using a "reasonable care" standard. Employers, including the controlling employer, must exercise reasonable care to prevent and detect violations on the site.

For example, in evaluating whether a controlling employer (typically a general contractor or builder) has exercised reasonable care in preventing and discovering violations, OSHA considers questions such as whether the controlling employer

- conducted periodic inspections of appropriate frequency
- implemented an effective system for promptly correcting hazards
- enforced the other employer's compliance with safety and health requirements with an effective, graduated system of enforcement and follow-up inspections

Factors that affect how frequently and closely a controlling employer must inspect to meet its standard of reasonable care include the following:

- project scale
- nature and pace of the work, including the frequency of hazard changes as the work progresses
- controlling employer's knowledge of the safety history and safety practices of the employer it controls and about that employer's level of expertise

More frequent inspections are normally needed if the controlling employer knows that the other employer has a history of noncompliance. Frequent inspections may also be needed at the beginning of the project if the controlling employer has never worked with the other employer and does not know its compliance history.

Less frequent inspections may be appropriate when the controlling employer recognizes strong indications that the other employer has implemented effective safety and health efforts. The most important indicator of an effective safety and health effort is a consistently high level of compliance. Other indicators include the use of an effective, graduated system of enforcement for noncompliance with safety and health requirements coupled with regular jobsite safety meetings and safety training. An example of this policy in action would be the following scenario:

ABC Builders hires Acme Plumbing and Main Electrical for a job. Acme and Main are working side-by-side on a structure. Main exposes Acme employees to electrical hazards, even though Acme's employees were instructed by Acme to stay clear of Main's work activities.

OSHA may cite all three parties because

- ABC Builders is the controlling employer
- Acme Plumbing is the correcting employer (a policy was in place to protect its employees from electrical hazards, but Acme did not enforce it)
- Main Electrical was the creating employer

Simply stated, the controlling employer—in many cases the builder—can be cited when other employers violate OSHA regulations on the jobsite. Courts have upheld this policy in the past. The scenario for a builder to be cited by OSHA for a trade contractor's noncompliance involves more than just the lack of posting the required emergency telephone numbers.

Some of the cases that courts have reviewed have extended beyond the relationship between the builder and trade contractor to the project manager and even engineering firms. A court held that a company performing construction management duties, but having no actual craft employees on site, was subject to the construction standards because of the supervisory role the company played on the jobsite. This ruling conjures up images of builders who subcontract their craftwork but have oversight of the construction activities. The direct parallels highlighted by the ruling are similar enough to concern many businesses.

Previous court decisions have also held the builder responsible for overall jobsite safety by virtue of the fact that the builder held supervisory capacity over the trade contractors on site. One of the best methods of avoiding OSHA citations for trade contractor violations is to only use trades that employ safety programs and policies that comply with OSHA standards. A safe, OSHA-compliant jobsite is good business for everyone. An injury-free jobsite can lead to decreases in workers' compensation rates and third-party liability lawsuits.

Preparing for an OSHA inspection can reduce your company's potential for citations. Preparation should include educating your workers about how to respond during an OSHA inspection and reinforcing your company's overall commitment to jobsite safety. Some of the nation's largest home builders demand that their trade contractors exhibit the same commitment to safety that they do and will not hire those who cannot meet their standards. This should be the goal for builders who are trying to protect their employees from injury and themselves from unwanted OSHA citations and costly injuries.

Model Safety Program for Builders

This chapter explains the components of the safety program. This program does more than just cover compliance with OSHA standards. It is the next generation of safety programs: a total loss-control program. This program is designed to be adaptable—not all of its sections are necessarily applicable to every builder.

This model safety program provides comprehensive guidelines to help your employees understand their responsibilities and to help you establish good safety procedures and create systems for collecting the information and records required by OSHA and your insurance company. It is based on the concept of total loss control. It centers on preproject planning to ensure that your safety program can be executed on every project to help you reduce or eliminate accidents that can cause

- injury to personnel
- damage to property
- fire or explosion
- hazards to health
- liability claims

The following comprehensive guidelines are provided in checklist format on the CD and in Appendix C for use in developing your own safety program. The model safety program is included in Appendix D and on the CD.

Components of a Total Loss-Control Safety Program

- **Safety and health.** The safety and health of your employees is your first priority.
- **Fire prevention.** Sound fire-prevention practices to eliminate potential fire hazards entail providing devices to detect potential fire hazards and to extinguish or control fires on the jobsite.
- **Property protection.** You can prevent property losses by controlling the working environment and your employees' actions.
- **Medical and first aid.** The scope of your jobsite medical/first aid program will vary by the number of employees, type of construction, potential health and safety exposures, and applicable OSHA standards.
- **Emergency planning.** Disasters can and do occur.

Steps for Developing a Comprehensive Safety Program

- State goals in writing
- Write an action plan
- Establish a budget
- Designate a "safety champion"
- Define areas of responsibility
- Develop and implement jobsite safe work practices
- Establish accountability procedures
- Develop a safety training program
- Develop enforcement procedures
- Conduct regular jobsite inspections and hazard analysis
- Develop recordkeeping procedures
- Establish accident reporting and investigation requirements
- Develop a trade contractor safety compliance policy
- Evaluate the effectiveness of your safety program
- Encourage feedback and reward excellence

STATE GOALS IN WRITING

What do you want to get from your safety program? After you determine your goals, you will have a great tool to build your program and measure its success. You might state that your company's goal is to provide the safest possible jobsite for employees as prescribed by the OSH Act. You might further state that the OSHA Construction Industry Safety and Health Standards (29 CFR 1926) are considered the minimum safety requirements for the company. Your company's project safety and loss-control goals can be succinctly stated as follows:

- protection of employees
- zero fatalities
- zero permanent disabilities
- prevention of injuries and illnesses
- prevention of any fires, vehicle accidents, or property damage

WRITE AN ACTION PLAN

Once you've defined your goals, develop an action plan to implement them. Your written plan should outline your commitment to run every project safely, comply fully with all federal and state safety laws and regulations, provide necessary equipment to ensure employee safety on the jobsite, and conduct ongoing education on safety practices for all employees.

ESTABLISH A BUDGET

Your budget must allow for training and safety equipment. In addition, you will need to factor in insurance costs, personnel to oversee the program, and time for safety meetings. By establishing a safety budget up front, you can set your safety program into motion.

DESIGNATE A "SAFETY CHAMPION"

You can develop a great safety program, but without oversight it will not be very effective. Select an employee who, in addition to his or her regular assignments, will be responsible for disseminating the company's safety message, implementing an effective safety program that also meets governmental requirements, and monitoring your program's results. Owners and project managers may not be the best choice for safety program oversight because production is often their primary focus. Make sure that your safety champion can focus his or her efforts where needed and that he or she has full support from management.

DEFINE AREAS OF RESPONSIBILITY

The following safety responsibilities may need to be adjusted to reflect your company's specific operations.

Company president/owner

Nothing can damage a safety program more than senior management breaking the rules. Therefore the president/owner leads by example when carrying out the following responsibilities.

The president/owner should

- provide direction and motivation and assign accountability to ensure an active safety and loss-control program for all company construction projects
- establish office and field administration safety and loss-control activities that reflect the company's safety goals and objectives
- establish a budget to fund the safety and loss-control programs

- develop annual safety goals and objectives for site superintendent(s) to meet as part of the superintendent's performance evaluations
- assist the site superintendent in developing site-specific safety and loss-control programs
- ensure that the management team has a working knowledge of all client, governmental, and company safety and loss-control requirements
- participate periodically in various employee safety toolbox presentations
- review monthly field safety status reports to evaluate each project's safety and insurance performance
- enforce incentive and disciplinary actions necessary to encourage an effective safety program

Site superintendent

Depending on the size of your company, some of the following duties may be split between the superintendent and other management personnel, such as a production manager or safety manager. Remodelers may find it more practical to assign the following duties to a lead carpenter or other lead employee.

The superintendent should

- establish comprehensive project safety procedures that comply with applicable client contractual documents and specifications (federal or state OSHA) and company safety and loss-control procedures with the assistance of the president/owner

- monitor the project's safety status and employee morale by conducting a daily safety inspection of the jobsite and initiating necessary corrective action
- conduct accident investigations, analyze the causes, and recommend corrective and/or preventive actions
- prepare accident reports and maintain documentation of workers' compensation reports
- maintain and update any necessary OSHA records and material safety data sheets (MSDS)
- conduct project-safety and loss-control training for employees
- ensure that each jobsite has the necessary safety equipment and materials, personal protective equipment, first aid supplies, and emergency telephone numbers posted
- monitor trade contractor performance to ensure compliance with the company's safety performance requirements
- prepare and distribute job safety bulletins and subject material for toolbox safety meetings and reviews and audit the meetings to ensure effectiveness
- enforce disciplinary actions
- notify the company's president/owner of any serious accident or OSHA inspection as soon as possible

Field employees

You should inform each employee that they are expected to work in a manner that will not inflict self-injury or cause injury to their coworkers.

All employees must

- comply with all safety rules and regulations
- immediately report all accidents and injuries to the supervisor
- use the proper tools and personal protective equipment for the job
- report all unsafe conditions to the supervisor
- know the procedures in the case of emergencies, including contacting emergency services
- help maintain a safe, clean work area
- participate in the company's safety training program
- set a good example for others to follow

Trade contractors

Your safety program should explicitly state that you expect trade contractors to have established their own safety and health programs. Each trade contractor is responsible for the safety of his or her employees on each company project.

Trade contractors must

- comply with applicable federal and state OSHA regulations

- supply a copy of their safety program and MSDS for all materials used on company projects
- immediately report all accidents, injuries, and fatalities that have occurred on company jobsite to the company superintendent
- supply the proper personal protective equipment and safety equipment to his or her employees and ensure their use
- train field employees on proper safety practices
- report all unsafe conditions to the site superintendent
- immediately notify the company president/owner or site superintendent in the event of an OSHA inspection when no company personnel are on site

DEVELOP AND IMPLEMENT JOBSITE SAFE WORK PRACTICES

Safe work practices are the key to the success of your safety program. Therefore, you must develop and implement safe work practices that will minimize the risks of potential hazards on your jobsites. Your safe work practices should include all applicable federal and state regulations, as well as practices that pertain to the specific conditions that are present on the jobsite. By performing a jobsite analysis, you can gain a clearer understanding of what those specific conditions may be. See Chapter 4 for a list of OSHA mandated and recommended safe work practices.

ESTABLISH ACCOUNTABILITY PROCEDURES

All employees must be aware of and vigorously pursue project safety goals. The company's interests must be vocal, visible, and continuous, so that all employees will know there is only one acceptable way to do the job—the safe way. You should inform every employee that they will be held accountable for safety and loss-control performance. This accountability will be reflected in retention, promotions, salary increases, and bonuses.

DEVELOP AND DELIVER A SAFETY TRAINING PROGRAM

You should properly train all employees (staff and management) in order for your safety program to be effective. This safety program can be used as the basis for your first training session. During the first training session, you should review the company's safety rules, detail how the rules will be enforced, and identify the staff person who is responsible for safety program oversight. You should offer safety training on an ongoing basis.

Supervisory training

You should conduct safety education through all phases of the work performed by the company. The site superintendent is responsible for the prevention of accidents at jobsites that are under his or her direction. The site superintendent is also responsible for

- conducting accident prevention and safety training

- supervising employees on the jobsite
- ensuring that jobsite employees comply with the applicable training provisions of OSHA regulations

Training records must be kept up-to-date and readily available for review during OSHA inspections.

Some builders have found that superintendents who have received basic first aid and CPR training are much more safety conscious and usually have better crew safety performance records. Each builder must decide whether to require his or her superintendents to be certified in first aid and CPR. (OSHA requires a jobsite to have at least one individual trained to render first aid in the absence of a treatment facility.)

Employee training

You should give a copy of the company's safety program and policies to all jobsite employees (temporary and full time) and require that each jobsite employee sign and date an acknowledgment of receipt. The acknowledgement can be kept in the employee's personnel file.

In addition, all employees should attend a training session that covers the company's job rules and regulations and the employee's personal safety requirements.

Toolbox safety talks

As the company president/owner, you should provide the direction and motivation to ensure that site superintendents conduct regular, brief worker safety meetings (15 minutes or less), also known as toolbox safety talks. The site superintendent should ensure that the discussion leader for each toolbox talk understands all company and site specific safety and loss-control policies and programs.

Interactive sessions tend to promote employee participation. You can also schedule toolbox safety talks at the beginning of new operations to ensure that all employees are familiar with safe work practices and the requirements of upcoming work.

Toolbox safety talks may be motivational or instructional in nature. The motivational meeting creates awareness and aims to achieve worker self-protection, whereas the instructional meeting covers a particular job task or procedure.

Emergency preparedness procedures

Telephone numbers for emergency service units should be posted on the jobsite. Site superintendents can render first aid and CPR, if qualified, until medical emergency personnel arrive. Site superintendents should know how to use fire extinguishers, guide employees to evacuation routes, and implement emergency procedures in the event of a fire. Training sessions for new hires should include the following information:

- emergency telephone number(s) for reporting a fire
- locations of fire alarm systems throughout the jobsite
- location and proper operation of fire extinguishers

- emergency evacuation routes and procedures

DEVELOP AND IMPLEMENT ENFORCEMENT PROCEDURES

It would be appropriate to issue a reprimand for the following reasons:

- failure to wear proper protective equipment, such as eye protection
- willfully endangering one's life or the lives of other employees, which is gross misconduct and will be cause for immediate dismissal
- performing work in an unsafe manner

Use the employee safety violation reprimand form (Appendix C) to warn an employee of an infraction of company safety rules prior to taking disciplinary action. The document should be kept in the employee's personnel file.

You can determine the severity of the discipline by the extent of the infraction. If the incident is the likely cause of an accident, or if the violation had a high probability of resulting in an accident, the employee may be terminated. If the incident had a moderate probability of causing an accident, time off without pay may be sufficient. If the incident had a low probability of causing an accident, the employee will receive a written reprimand. The site superintendent should personally advise the employee that three written reprimands for safety violations will result in immediate termination. Most companies follow a three-step warning system:

Step 1: Employee receives a verbal warning about his or her actions and is expected to comply immediately.

Step 2: If the employee is found in violation a second time or the employee has not corrected the behavior for which he or she received a verbal warning, then a written violation notice is given to the employee. In multiemployer situations, a copy of the violation is also sent to the employee's employer and another copy is kept on the jobsite.

Step 3: If an employee is discovered in violation a third time or the employee still has not corrected the behavior that resulted in the previous two warnings, then the employee can be removed from the project.

The most important aspect of disciplinary action is to make sure that it is fairly and consistently imposed.

CONDUCT REGULAR JOBSITE INSPECTIONS AND HAZARD ANALYSIS

Jobsite inspections allow employers to gauge the effectiveness of their safety programs and determine employee training needs. Although these inspections are required by OSHA, this critical step of jobsite safety is often overlooked.

The site superintendent should conduct a weekly jobsite safety audit and complete and submit a signed construction safety checklist (Appendix C) to the president/owner. This responsibility should not be delegated to other staff members. As part of an effective jobsite inspection program, the site superintendent must

- set inspection responsibilities and schedules
- develop an administrative system to review reports
- devise follow-up procedures for corrected conditions
- analyze inspection findings
- set program standards for observing employee safety practices
- communicate program standards for observing employee practices to each supervisor
- communicate program safety standards to employees
- monitor employee safety practice performance

You should inspect the jobsite with the site superintendent to review working conditions and ensure that the sites are in compliance with the company's safety policies. You and the site superintendent should discuss the status of site safety and loss-control programs and performance results to date, as measured against the company's targeted goals on every jobsite visit.

Carriers of the company's workers' compensation, general liability, and automobile insurance may also need to conduct a jobsite safety inspection or accident investigation. As the president/owner you should approve these scheduled insurance safety audits and notify the site superintendent when the insurance representatives will be on site. Site supervisory personnel should fully cooperate with the company's insurance representatives.

Jobsite hazard analysis

It is important to include a jobsite hazard analysis (JHA) in your safety program. By performing a JHA, you can proactively address potential hazards and take corrective measures. You can also implement a safe work practice that will not only reduce hazards, but can be used to train employees on how to protect themselves in the future.

Many large companies employ full-time safety managers to check each site and prepare checklists for the foreman and superintendents to use to bring trade contractors into compliance. Smaller companies tend to use consultants, who can be hired on a part-time basis, to perform the same function. Risk managers from insurance companies often perform inspections, if requested, a few times a year. However, risk managers generally inspect sites on a quarterly basis, and a lot goes on between inspections.

Proactive Hazard Analysis

When ABC Contracting performed a jobsite hazard analysis, they found an area where employees would be exposed to a 14-ft. trench. The problem was that there was no way to slope the type "C" soil to meet OSHA standards. To address the potential hazard of a trench collapse, ABC rented a shoring system that was delivered a day before excavation began. This proactive approach prevented a project slowdown and protected employees from a potential trench collapse.

DEVELOP RECORDKEEPING PROCEDURES

Various accident and injury reports and records are necessary to meet the requirements of the company, insurance carriers, and government regulatory agencies. These uniform procedures apply to all company jobsites and are used to measure the overall safety and insurance performance of each company project. The site superintendent can delegate the daily administration of these reporting and recordkeeping requirements to a staff member. In that event, however, the site superintendent should determine the actual timely and adequate completion and distribution of these reports and records.

Copies of forms and records will not be duplicated or distributed to unauthorized personnel, outside agencies, employees, or other third parties without the explicit permission of the president/owner. Requests for forms or records from third parties or external agencies should be approved by the president/owner. This includes requests from clients and owners of projects.

Forms devised for use at field locations should be approved by the president/owner prior to their use. Any and all records generated at field locations must be maintained at the location until completion of the project. Safety or medical files or records must not be destroyed.

Company records

In addition to workers' compensation reports, the site superintendent should maintain a file of all company safety records, including accident investigation reports, construction safety checklists, and the safety orientation checklist for trade contractors (Appendix C).

OSHA Log of Work-Related Illnesses and Injuries (Form 300)

This log must be maintained by employers who have employed more than 10 employees during the previous calendar year. "Employee" does not include trade contractors, who are considered separate employers by OSHA and who must maintain their own OSHA records.

The site superintendent is responsible for maintaining the log of accidents and injuries. General instructions for completing the log are included on the reverse side of the form. The site superintendent is responsible for completing and signing the log. The log must be retained for five years. In addition, the following conditions must be observed:

- When company work from the previous year is still ongoing, the *Summary of Work-Related Illnesses and Injuries* (Form 300A) must be posted on the company jobsite bulletin from February 1 to April 30, after which it may be taken down and filed with other jobsite safety records.

- Under no circumstances should the company site superintendent maintain a log for trade contractors.

Form 300 and 300A can be downloaded from www.osha.gov/recordkeeping/newosha300form1-1-04.pdf.

Reporting fatalities and multiple hospitalizations

Within eight hours after the death of any employee from a work-related incident or the in-patient hospitalization of three or more employees as a result of a work-related incident/accident, you must verbally report the fatality/multiple hospitalization to the OSHA area office that is closest to the site of the incident by phone or in person (see Appendix B). You may also use the OSHA toll-free number 1-800-321-OSHA (1-800-321-6742).

ESTABLISH ACCIDENT REPORTING AND INVESTIGATION REQUIREMENTS

The purpose of accident investigation is not to assign fault but to understand what led to the event so that you can protect company employees in the future. The information can also be used to help determine future training needs so that similar accidents and near misses can be prevented.

Near misses are events in which damage to individuals and/or property could have occurred. For example, two employees were working in a trench leading to a house. The trench was 6 ft. deep and not shored or sloped. The soil was type "C," and there were no ladders in the trench. The employees stopped for lunch, and when they returned, the trench had collapsed.

Accidents are events that cause actual damage to individuals and/or property. For example, two employees died in a trench collapse in Potomac, Maryland, on July 19, 2006. The men were applying waterproofing material to the foundation of a home when the trench caved in, and both men were buried under the dirt.

Every person in the company, including employees, managers, and owners, is responsible for investigating accidents. However, the site superintendent's unique position gives him or her additional responsibility in this area. The superintendent usually knows the most about an accident because of his or her familiarity with the jobsite and can take the most immediate action to prevent an accident from recurring. The site superintendent is also able to effectively communicate with employees on site.

The company considers an accident to be serious if it results in

- occupational death(s), regardless of the time between injury or illness and death
- occupational injuries or illnesses resulting in permanent total disabilities
- occupational accident(s) that involve any property damage
- hospitalizations

Site superintendents should complete an accident investigation report form (Appendix C) after an accident has occurred. The form meets OSHA's recordkeeping requirements for recordable accidents and it allows superintendents to assemble valuable data that may be used to prevent accidents on future projects.

All statements about any accident made to individuals not connected with the company should be handled by the president/owner. Statements that must be made by company field personnel to insurance company representatives or law enforcement authorities should be confined to the "basic facts." Further details should be cleared by the company president/owner prior to their release. No statement regarding accident *liability* should be made to anyone not connected with the company.

Conducting the investigation

Employees should be trained to conduct an accident investigation and have the necessary materials, or an event kit, to do so. An event kit could contain the following items:

- paper
- pens
- accident investigation report form
- camera
- tape measure
- flashlight
- audio/video recorder

In order to collect the most accurate witness statements, investigations should be conducted immediately after help has been rendered and the jobsite is safe. OSHA requires employers to keep records regarding accidents on the jobsite. The information pertaining to any incident requiring medical treatment beyond first-aid must be recorded on the OSHA 300 form.

Corrective action must take place immediately after the investigation to ensure that the event does not recur. You should analyze the report and develop a timeline to help implement any changes that need to be made.

DEVELOP AND MAINTAIN A TRADE CONTRACTOR SAFETY COMPLIANCE POLICY

All trade contractors, regardless of the size of the company, must comply with all local, state, and federal regulations, including, but not limited to, OSHA's minimum rules and regulations and the builder's safety policy. For example, if a trade contractor is working with a builder who requires that all employees have proper fall protection above 4 ft. in height, then the contractor must meet both OSHA's standard requirements for scaffold fall protection and the builder's standards in order to be compliant.

Criteria for choosing safe trade contractors

Trade contractors who employ safe work practices generally tend to produce a high-quality product in an organized, efficient manner. Although these contractors may have slightly higher rates, many builders agree that working with safety-conscious trade contractors ultimately saves time and money over the life of the project. Not only does working with safety-conscious contractors allow you to produce a quality product, but you can expect fewer accidents and a decrease in the likelihood of an OSHA inspection. Use the following criteria to evaluate a trade contractor.

- Do they have a safety program?
- When was the program last reviewed?
- Who is in charge of the program?
- How do they ensure that the safety program is being followed?
- How often are inspections made?
- Who would be responsible for safety for your project?
- How do they enforce procedures?
- Have their employees received safety training?
- When was the last training session offered?
- Do they provide ongoing training?

Preconstruction safety orientation

It is very important to outline all requirements for each job prior to beginning the work. Preconstruction safety orientation meetings allow you to explain your safety program and disciplinary policies, so that trade contractors are aware of the consequences. You can also use this meeting to discuss any site-specific rules, including work hours and required personal protective equipment, as well as address any questions or concerns.

EVALUATE THE EFFECTIVENESS OF YOUR SAFETY PROGRAM

The number of accidents that have occurred should not be the only factor considered when determining program effectiveness. Your evaluation should include a review of any written programs, training records, jobsite inspections, injuries/illness investigations, and the OSHA 300 Form. Each of these items can help you identify potential deficiencies in the overall program. If your safety program is not working effectively, you might want to review the quality of your training sessions or improve your safe work practices.

ENCOURAGE FEEDBACK AND REWARD EXCELLENCE

You should periodically review and revise your safety program. Employee feedback can be a valuable resource. Be sure to seek and include their suggestions for improvement. Let your employees know that you recognize and appreciate their commitment to safety through ongoing positive reinforcement. This reinforcement may take the form of cash, compensatory days for a given number of accident-free days, or special recognition in the workplace. A daily word of encouragement or a suggestion on how to do a job more safely raises your employees' awareness of safety's importance to you and your company.

OSHA Regulations and Safe Work Practices

This chapter provides comprehensive safe work practices and OSHA requirements for residential contractors. Your safety program should include the safe work practices for the conditions on your jobsites. These practices are included on the CD that accompanies this book for ease in incorporating into your safety program.

Since the Occupational Safety and Health (OSH) Act was enacted in 1970, employers engaged in residential construction have been subjected to OSHA requirements. In the last several years, OSHA has begun to focus its enforcement efforts in the residential sector of the construction industry. As a result, builders have had to be diligent in identifying the regulations that apply to their businesses in order to ensure that they are in compliance.

This chapter presents the OSHA requirements that most residential contractors must comply with and outlines both employer duties and employee responsibilities. The information contained in this chapter should be used as a guideline. Following these guidelines does not exempt you from compliance with the requirements in Title 29 Code of Federal Regulations Part 1926, any state or local safety laws and regulations, or applicable standards for the residential construction industry.

PERSONAL PROTECTIVE EQUIPMENT

Requiring employees to wear personal protective equipment (PPE) is one of the easiest ways to protect them from many of the hazards found on residential construction jobsites. OSHA requires the following types of PPE on residential jobsites:

- hard hats
- eye protection
- hearing protection
- work shoes or boots
- gloves
- respirators
- safety glasses

Safety-toe footwear must meet the American National Standards Institute's (ANSI) Z41 requirements for protection. Check the tongue of the boot for the Z41 label. Safety glasses that are required for impact protection must meet the ANSI Z87 requirements. If your employees wear vision-correcting glasses, these must meet the ANSI standard too. If not, the employee must wear the protective glasses over the vision-correcting glasses. Check for the Z87 stamp to ensure that the glasses are impact-resistant safety eyewear.

Providing PPE should be your last resort for employee hazard protection. Your primary goal should be to remove the hazard or to implement control measures that limit employee exposure. For example, if your employees are working in a confined area with a material that can be harmful if inhaled, you should first try to remove the hazard by ventilating the area. If that is not possible, then you should use a less hazardous material. If that option is not feasible, then you should require your employees to wear the proper protection, in this case, a respirator.

Employer responsibilities

You must provide the appropriate PPE for your employees whenever there is a hazard that cannot be eliminated through engineering or administrative controls. You must also require employees to follow safe work practices.

OSHA requires that you have a written respiratory protection program when respiratory protection is needed for employee safety. You should refer to the full text of OSHA's 1926.103 standard for more information. Review OSHA 1926.101 if you need hearing protection to protect workers on the jobsite.

It is important that you establish specific rules for the wearing of PPE on the jobsite. You must be consistent in the enforcing these rules. Many builders have programs such as the "100% hard hat" rule that requires any person who comes onto the jobsite to wear a hard hat all the time. It is important that you also follow the rules. Therefore, if hard hats are required, then you should always be wearing one on site too.

Maintenance

If the employer owns the PPE, OSHA requires that it be maintained in a reliable and sanitary condition that includes having the equipment cleaned and disinfected as needed. If the employee owns the PPE, the employer still must ensure that it is being properly used and that it meets OSHA's requirements.

Training requirements

All employees must be trained in proper PPE use. Employers are also required to provide training for employees who must use respirators. The training must be comprehensive and understandable and must be offered on at least an annual basis.

Safe work practices

All employees must adhere to the following safe work practices.

- Wear a hard hat when there is a danger from impact, falling or flying objects, or electrical shock.
- Wear impact-resistant safety glasses when you use materials or operate equipment that could result in materials striking your eyes.
- Wear safety goggles if you are working with materials or chemicals that could damage your eyes on contact.
- Wear face shields to protect your face from flying objects or splash hazards.
- Wear proper eye protection when welding, cutting, or brazing.

- Use hearing protection when exposed to hazardous levels of sound.
- Wear proper shoes or boots while on the jobsite to protect against nail puncture injuries.
- Wear respiratory protection when you are exposed to inhalation hazards.

FIRE PROTECTION AND PREVENTION

Fires are common on residential construction jobsites because employees are often working with hazardous materials, doing hot work, and using temporary heating devices. Fires can result in serious damage to property, serious injuries, and death. Properly scheduling your trade contractors is one way to limit the fire hazards. Think about who has to be on the jobsite at the same time. Consider this formula for disaster—a plumber is doing hot work while another trade is using a flammable material nearby.

The four classes of fires are

- Class A—paper, wood, straw, cloth
- Class B—flammable and combustible liquids
- Class C—energized electrical equipment
- Class D—combustible metals

Testing laboratories classify fire extinguishers based on the class of fire they are designed to extinguish:

- Class A fires can be extinguished with water.
- Class B fires can be extinguished with carbon dioxide, foam, or dry chemical.
- Class C fires can be extinguished with carbon dioxide or dry chemical.
- Class D fires can be extinguished with special extinguishing compounds.

Employer responsibilities

You must develop a fire protection program that includes the necessary fire-fighting equipment and a system to warn employees and notify the fire department. Use the following steps as a guideline for implementing a fire protection program:

- Provide at least one 2A (or better) fire extinguisher for each house under construction (water drums and hoses are acceptable alternatives in the absence of fire extinguishers, if they meet the OSHA requirements).
- Keep access to fire extinguishers clear.
- Train workers on how to operate fire extinguishers.
- Inspect fire extinguishers periodically.
- Develop a system to notify employees of a fire on site.
- Develop a system to account for all employees on site.
- Post fire notification procedures where employees can find them.

Fire prevention

Be sure that electrical wiring and equipment are installed and maintained according to OSHA requirements and that any internal combustion engines are located away from flammable materials. Smoking must not be allowed within 50 ft. of fire hazards, and signs must be posted to indicate that smoking and open flames are not allowed.

Open-yard storage

Proper storage of materials on the jobsite can help keep fire hazards to a minimum. Do not store combustible materials more than 20 ft. high, and keep driveways between and around combustible material piles at least 15 ft. wide. Maintain good housekeeping, and keep areas clear of debris, weeds, and grass.

Flammable and combustible materials storage and handling

Use approved containers for storing and handling flammable and combustible liquids. Use approved metal safety cans for handling any flammable liquids of more than one gallon, unless it is a viscid (extremely hard to pour) material; in that case, the original container is acceptable.

Never store flammable or combustible liquids near doors or exits. If you are storing more than 25 gallons of a flammable liquid inside a building, it must be kept in an approved storage cabinet labeled "Flammable—Keep Fire Away." Do not store more than 60 gallons of flammable liquids in a single approved storage cabinet.

No more than 120 gallons of combustible liquids are permitted to be kept in a single approved storage cabinet, and they must always be kept at least 10 ft. from a building or structure.

Containers of flammable liquids must be kept closed when not in use. If a spill occurs, employees must clean it up immediately and properly dispose of the spilled liquid. You must also ensure that there is no smoking or open flame within 50 ft. when flammable liquids are being used.

Temporary heating devices

If you use temporary heating devices on your jobsites, be sure that there is enough air circulation to maintain proper combustion and limited temperature rise. Portable heaters must be placed on heat-insulated material that extends 2 ft. or more beyond the heater in all directions. Keep heaters level and follow the manufacturer's specifications. Keep all tarpaulins and similar materials at least 10 ft. away from the heater. Be sure employees never use solid fuel salamanders in buildings or on scaffolds.

Potential ignition sources

The residential construction industry uses a variety of materials and products that are potential ignition sources. Be sure your employees know which materials are flammable and what precautions they need to take to avoid fires on the jobsite. If you burn waste on the site, be sure that the fire is maintained and that employees do not burn hazardous materials. Materials likely to be fire hazards include:

- gasoline
- liquid nails or similar adhesives
- form oil
- some paints, solvents, and mastics

Training requirements

Employees must be trained to do the following:

- Ensure that everyone has left, or is leaving the structure and is accounted for once evacuated.
- Call the fire department.
- Ensure that the fire is confined to a small area and is not spreading.
- Find an unobstructed escape route.
- Know how to properly operate the fire extinguisher.

Portable fire extinguishers must be inspected periodically and maintained in accordance with Maintenance and Use of Portable Fire Extinguishers, NFPA No. 10A-1970. ANSI Standard 10A-1970 states, "The name plate(s) and instruction manual should be read and thoroughly understood by individuals who may be expected to use extinguishers." To use a fire extinguisher properly, employees should be trained in the PASS procedure:

- **P**ull the pin and break the seal.
- **A**im the nozzle at the base of the flame.
- **S**queeze the lever while holding the extinguisher upright.
- **S**weep the nozzle from side to side.

Safe work practices

All employees must adhere to the following safe work practices.

- Know where the fire extinguishers are located and how to use them.
- Use only approved safety cans for storing more than one gallon of flammable liquid, although the original container may be used for less than one gallon.
- Do not store flammable or combustible liquids in areas used for stairways or exits.
- Do not store combustible materials more than 20 ft. high.
- Keep driveways between and around combustible material piles at least 15 ft. wide.
- Keep areas clean of debris, weeds, and grass.
- Do not store combustible materials within 10 ft. of a building or structure.
- Keep fire extinguishers with a rating of at least 2A within 100 ft. of storage areas.
- Keep fire extinguishers in plain sight and within reach.
- Store flammable liquids in closed containers when not in use.
- Clean up leaks or spills of flammable or combustible liquids promptly.
- Use flammable liquids only where there are no open flames or other ignition sources within 50 ft. of the operation.
- Do not store liquefied petroleum (LP) gas tanks inside buildings.
- Keep LP gas containers with a water capacity greater than 2½ lb. on a firm and level surface and, when necessary, in a secured, upright position.

- Keep temporary heaters at least 6 ft. away from any LP gas container.
- Do not use solid fuel salamanders in buildings or on scaffolds.

TOOLS

Tools are a very important part of the residential construction industry. However, when used incorrectly, tools can cause accidents, even for the experienced construction worker.

In addition to training, you must use proper guarding and keep tools maintained in order to protect your employees. Cuts, bruises, and even death can be the unfortunate results of removing guards or bypassing safety devices on tools.

Employer responsibilities

All hand and power tools, whether they are owned by the employer or employee, must be maintained in a safe condition. Employers must never issue or permit the use of any unsafe tool, and employees must receive specific training on powder-actuated and woodworking tools before using them.

Tools must be grounded or be double insulated. Any tool that is identified as damaged or defective must be immediately removed from service.

You must ensure that all guards are in place for tools that require protection. Any reciprocating, rotating, or moving parts that could come into contact with employees must be guarded; for example, belts, gears, shafts, pulleys, sprockets, spindles, drums, flywheels, and chains.

In addition, you must guard all air hoses exceeding 1/2 in. in inside diameter with a safety device at the source of the supply or branch line to reduce pressure in case of hose failure. You must guard airless spray guns that atomize paints and fluids at 1,000 psi or greater with a safety device to prevent the release of paint or fluid, or use a diffuser nut to prevent high-pressure and high-velocity release when the nozzle tip is removed. Employers must also provide a nozzle tip guard to prevent the tip from coming into contact with the employee.

Training requirements

Employees must be trained to operate powder-actuated tool before using these tools. All woodworking tools and machinery shall meet other applicable requirements of the ANSI Standard 01.1-1961. ANSI Standard 01.1-1961 states, "Before a worker is permitted to operate any woodworking machine, he shall receive instructions in the hazards of the machine and the safe method of its operation."

Safe work practices

All employees must adhere to the following safe work practices.

General guidelines

- Maintain all hand and power tools (employee or employer owned) in safe condition.
- Follow the manufacturer's requirements for safe use of all tools.
- Inspect all hand and power tools before use. If a tool is found to be defective,

OSHA REGULATIONS AND SAFE WORK PRACTICES

remove it from service and notify the supervisor.

- Keep guards on tools at all times.
- Use all required personal protective equipment when using tools.
- Do not use wrenches when the jaws are sprung to the point of slippage.
- Do not use impact tools with mushroomed heads.
- Keep wooden handles free of splinters and cracks and ensure that they fit securely in the tool.
- Never point tools at anyone.
- Test the safety device on the tool each day before using.
- Do not leave loaded tools unattended.
- Use personal protective equipment that meets OSHA requirements.
- Ask your supervisor for help or training if you need it.

Power tools

- Keep all electric power tools grounded or use the double-insulated type.
- Do not use electric cords or hoses for hoisting or lowering tools.
- Secure pneumatic power tools to the hose or whip to prevent them from accidentally disconnecting.
- Use safety clips or retainers on pneumatic impact tools to prevent attachments from accidentally expelling.
- Do not use compressed air for cleaning, except when below 30 psi and only with personal protective equipment. The 30-psi requirement does not apply to concrete form, mill scale, and similar cleaning purposes.
- Work within the manufacturer's safe operating pressure for hoses, pipes, valves, filters, and other fittings.

Woodworking tools

- Use only portable, power-driven circular saws that have guards above and below the base plate or shoe. When the tool is withdrawn from work, the lower guard must automatically and instantly return to the covering position.
- Maintain guards on all the moving parts and blades of other portable saws and equipment.
- Load tools only just before use.
- Do not fire into materials that are too hard or too soft.
- Reinforce materials if there is a chance that fasteners could go through the material.
- Do not try to fasten an area that is spoiled from other attempts to fasten.

Powder-actuated tools

- Only operate a powder-actuated tool if you have been trained in the operation of the particular tool.
- Test the tool before loading each day in accordance with the manufacturer's recommended procedure to ensure that the safety devices are in proper working condition.

- Do not load tools until just before the intended firing time.
- Keep hands clear of the open barrel end.
- Do not drive fasteners into very hard or brittle material.

WELDING AND CUTTING

Soldering, welding, or cutting can cause many hazards on residential construction jobsites. Welding operations can be performed safely if reasonable precautions are taken. However, handling compressed gas cylinders and working with acetylene can be very hazardous if employees have not been properly instructed in safe work practices. If the neck of a pressurized cylinder accidentally breaks off, the force could cause the cylinder to be launched and cause serious harm to people and property.

Employer responsibilities

It is important for employers to understand that in order to create a safe environment for working with compressed gases, employees must be provided with the proper equipment, supplies, and training.

Wood floors should be swept clean and covered with a non-flammable material before workers weld over them. In some cases, it might be advisable to wet the floor down. If you choose this method to ensure safety, you must also guard against the added shock hazard if using electrical welding equipment.

Fire protection must always be available during welding operations. A portable fire extinguisher must be kept at hand, especially if there are any combustible materials in the area.

Good ventilation is a must in any welding operation. Many welding operations produce fumes that can be harmful in heavy concentrations. In certain situations, it may be necessary to provide special ventilating equipment, such as portable exhaust fans.

PPE is always required for welding operations. The proper PPE includes eye and face protection and protective clothing. If the rules for PPE are followed, the risk of injury is minimized.

Training requirements

You must train employees in the safe use of fuel gas and the safe means of arc welding and cutting.

Safe work practices

All employees must adhere to the following safe work practices.

Transporting, moving, and storing compressed gas cylinders

- Protect valves of stored cylinders.
- Secure cylinders on a cradle, sling board, or pallet when hoisting.
- Do not hoist cylinders with magnets or choker slings.
- Move cylinders short distances by tilting and rolling them on the bottom edges, not the sides.
- Do not drop cylinders or allow them to collide.

- Secure cylinders in an upright position during transport.
- Do not lift cylinders by the valve caps.
- Do not use bars under valves or valve caps to pry cylinders loose when frozen.
- Secure valve caps before moving cylinders unless they are secure in a cylinder carrier.
- Secure cylinders in a truck or cart to prevent them from falling or being knocked over while in use.
- Do not use cylinders as rollers or supports.
- Close all valves before moving cylinders.
- Secure cylinders in an upright position except when briefly hoisting or carrying them.
- Separate oxygen cylinders in storage from fuel/gas cylinders or combustible materials cylinders (especially oil or grease). This separation must be either a minimum distance of 20 ft. or a non-combustible barrier at least 5 ft. high with a ½-hour fire-resistance rating.
- Store cylinders in a well-protected, well-ventilated, dry location at least 20 ft. from highly combustible materials such as oil or excelsior.
- Do not store cylinders in unventilated enclosures such as lockers and cupboards.
- Protect stored cylinders from tampering and damage.

Usage

- Crack the valve before connecting a regulator by opening the valve slightly and then immediately closing it.
- Stand on the side of outlets when cracking the valve.
- Do not crack fuel gas cylinders near welding work, sparks, flame, or other ignition sources.
- Open the valve slowly to prevent damage to the regulator.
- Do not open quick-closing valves by more than 1½ turns.
- Leave any needed wrenches in position on the stem of the valve while the cylinder is in use so that the fuel gas flow can be shut off quickly if necessary.
- Do not place anything on top of a cylinder in use that may damage the safety device or interfere with the quick closing of the valve.
- Do not use fuel gas from cylinders through torches or other devices without a pressure-reduction regulator attached to the cylinder valve.
- Close the valve and release gas from the regulator before removing the regulator from the cylinder valve.
- Close the valve and tighten the gland nut of fuel gas cylinders when a leak is found around the valve stem.
  - If this action does not stop the leak, disconnect the cylinder and remove it from service, tag it, and remove it from the work area.

- If fuel gas leaks from the cylinder valve rather than from the valve stem and the gas cannot be shut off, tag the cylinder and remove it from the work area.
- If a regulator attached to a cylinder valve stops a leak through the valve seat, the cylinder does not need to be removed from the work area.
- If a leak develops at a fuse plug or other safety device, remove the cylinder from the work area.

- Use fire shields or keep cylinders far enough away from welding or cutting operations so that sparks, hot slag, or flame cannot reach them.
- Keep cylinders away from energized electrical circuits.
- Do not strike a cylinder with an electrode to strike an arc.
- Use fuel gas cylinders in an upright position only.
- Keep fuel gas cylinders away from flame, hot metal, or other sources of heat.
- Keep cylinders containing oxygen, acetylene, or other fuel gas out of confined spaces.
- Keep a fire extinguisher in the work area of welding, cutting, or flames.
- Remove damaged or defective cylinders from the work place.
- Do not use oxygen for ventilation, comfort cooling, blowing dust from clothing, or for cleaning work areas.
- Keep oxygen cylinders and fittings away from oil and grease.
- Keep cylinders, caps and valves, couplings, regulator hoses, and apparatus free of oil or greasy materials and don't handle them with oily hands or gloves.
- Maintain ventilation during operations.
- Pressure regulators and gauges must be in proper working order while in use.
- Use fans, open windows, or work outdoors to keep fume levels low.

Hoses
- Make sure you can differentiate between fuel gas and oxygen hoses. Use different colors or surface characteristics. Oxygen and fuel gas hoses are not interchangeable.
- Do not use a single hose that has more than one gas passage.
- Do not cover more than 4 of 12 in. when taping parallel sections of oxygen and fuel gas hose together.
- Inspect all hoses that will carry any gas or substance that may ignite or be harmful.
- Remove defective and damaged hoses from service.
- Use double-locking or rotary connections for hoses.
- Ventilate storage areas and boxes used to store gas hoses.
- Keep passageways, ladders, and stairs clear of hoses and other equipment.

Torches

- Clean clogged torch tip openings with suitable cleaning wires or drills.
- Inspect torches before use for leaking shutoff valves, hose couplings, and tip connections.
- Do not use defective torches.
- Use friction lighters to light torches.

ELECTRICAL SAFETY

Electrocution is one of the top four causes of fatalities on construction sites. Recently, OSHA has focused more effort on inspecting jobsites for electrical hazards. These hazards range from worn extension cords to potential contact with overhead and buried power lines. Electricity poses many dangers:

- It can burn. The burns can be internal or surface. Electrical current can heat jewelry or tools and burn the wearer or operator. An electric arc flashing through the air can also burn.
- It can shock. An electrical shock can stop a person's heart, cause a breathing cessation, and muscle contraction that could result in a fall and further injury.
- It can cause a fire or an explosion.

Electricity is used so frequently on residential construction sites that the hazards it poses no longer seem threatening. But this type of thinking can lead to accidents, injuries, and death. It is very important that you properly maintain your electrical equipment. It is imperative that you enforce the employee safe work practices listed at the end of this section.

Employer responsibilities

Ensure that all electrical equipment used on site is hazard-free, in terms of:

- suitability of the equipment
- durability
- mechanical strength
- insulation
- heating effects
- arcing effects
- classification of use

Conduct regular inspections of electrical equipment to ensure that all defective or damaged equipment is removed from service for repair or destruction. Never use faulty equipment!

You should conduct regular jobsite inspections to determine the potential for employee exposure to electrical current due to missing electrical covers, knockouts, or enclosures. Any exposures noted should be corrected immediately.

Make sure the electrician maintains all electrical equipment per the National Electrical Code (NEC) for grounding, Ground Fault Circuit Interrupter (GFCI) placement and use, equipment placement clearances, and temporary power. All temporary cabinets and boxes must be protected from moisture buildup during the period that the job is ongoing. Outdoor circuit breakers and switches must be installed in weatherproof boxes.

You should only allow hard service or junior hard service extension cords to be used on site. All casing on the cords must be marked with the following letters: S, ST, SO, STO, SJ, SJT, SJO, or SJTO. Flat cords must never be used on a construction site. You must make sure that all temporary 120-volt, single-phase, 15- and 20-ampere receptacle outlets (generators, extension cords, and temporary power poles) are protected with approved GFCIs.

Employees must not work near live electrical parts unless they are protected from shock either by grounding or some other effective means. Employers must implement a lockout/tag out system to prevent electrocution.

Training requirements

OSHA requires that employers "Instruct each employee in the recognition and avoidance of unsafe conditions and the regulations applicable to his work environment to control or eliminate any hazards or other exposure to illness or injury."

Safe work practices

All employees must adhere to the following safe work practices:

General

- Do not contact any electrical power circuit unless the circuit is de-energized or guarded by insulation or other means.
- Wear insulated gloves when using jackhammers, bars, or other hand tools that may contact a line when the underground location of the power lines is unknown.
- Remove all damaged electrical tools from service.
- Protect electrical equipment from contact in passageways.
- Keep all walking/working surfaces free of electrical cords.
- Do not use worn or frayed electrical cords or cables.
- Do not fasten extension cords with staples
- Do not hang cords from nails or suspend them with wire.
- Maintain a minimum of 10 ft. from all energized power lines.

Ground fault protection and temporary power

- Use GFCIs to protect extension cords and any other connectors even if the cords are connected to the permanent wiring of the house.
- Install all temporary receptacles in complete metallic raceways.
- Protect all general lighting lamps from breakage.
- Ground all metal case sockets.
- Protect extension cords that are run through doors, windows, and floor holes.
- Use only three-wire type extension cords designed for hard or junior hard

service. Look for the following letters imprinted on the casing: S, ST, SO, STO, SJ, SJT, SJO, or SJTO.

- Do not bypass any protective system or device designed to protect you from contact with electrical current.

Lockout/tag out
- Controls that are to be deactivated during the course of work on energized or de-energized equipment or circuits must be tagged or marked.
- Equipment or circuits that are de-energized must be rendered inoperative.
- Attach tags at all points where equipment or circuits can be energized.
- The tags should be placed to clearly identify which pieces of equipment or circuits are being serviced.
- The lock or tag can only be removed by the employee who applied it.

SCAFFOLDS

The residential construction industry uses a wide variety of scaffolds. Because scaffolds are only used temporarily, they are seldom constructed as well as more permanent structures. As a result, over one-third of the serious injuries are caused by falls from an elevated working level, usually because the employee did not have a safe place to stand while working.

A recent OSHA analysis of injuries and fatalities relating to scaffold use indicated the following:

- 72% of employees injured in scaffold accidents attributed the accident to the planking/support giving way, slipping, or being struck by a falling object.
- 70% of the employees learned the safety requirements for installing work platforms, assembling scaffolds, and inspecting scaffolds through on-the-job training. Approximately 25% of the employees had no training in these areas.
- Only 33% of scaffolds were equipped with guardrails.

OSHA also indicated that scaffold-related fatalities account for approximately 9% of all fatalities in the construction industry.

Although erecting scaffolding requires some effort, they should be designed and built according to safety rules. One of the most important elements of scaffold construction is to ensure that the scaffold components are designed to work together. OSHA does not permit builders to intermingle scaffold components from different manufacturers unless the competent person has determined that the components will fit together without force and the scaffold is structurally sound.

A safe scaffold structure requires a firm foundation. In order to avoid settling, all poles, legs, posts, frames, and uprights must be placed on base plates and mud sills, or some other firm foundation, where necessary. Unstable objects such as scrap lumber, terra cotta, or concrete block fragments must not be used to support scaffolds.

It is important to follow the safety rules for scaffolds and scaffold platforms. Unstable or unsafe objects such as plywood, old or decaying lumber, or siding panels must never be used as working platforms.

Employer responsibilities

You are responsible for properly training your employees on the safe erection and use of scaffolds. All employees who are involved in the erection, use, or dismantling of scaffolds must follow proper safety practices. Serious injuries and even fatalities can result when scaffolds are improperly erected, used, or dismantled. In addition, there are some very specific requirements for the employee who is deemed to be the competent person in scaffold work. You must designate a competent person to oversee safe scaffold work practices on your jobsites. This person can be a current employee who has demonstrated experience in identifying and correcting potential scaffold hazards.

The competent person should be responsible for performing an initial evaluation of the jobsite in order to design and plan the job. OSHA has specific training requirements for the competent person.

In the residential construction industry, no two jobsites are exactly alike; therefore careful judgment must be used to prevent injuries. Some of the areas to be evaluated are as follows:

- proximity to electrical lines
- adequate access to the jobsite
- weather conditions
- ground conditions
- foundations of sufficient strength to support scaffolds
- interference with other work being performed
- environmental hazards
- high traffic area

Improper or makeshift scaffolds create an unsafe work environment. The following guidelines are included to help you comply with the requirements of the scaffold standard. You can use these guidelines as a starting point for designing your scaffold systems. The guidelines do not provide all the information necessary to build a complete system. Therefore you are still responsible for designing and assembling scaffold components in a manner that will meet the requirements of the OSHA standard. The competent person must ensure that all OSHA requirements are followed.

Single and double (or independent) pole

Single-pole scaffolds consist of a platform that rests on bearers. The outside ends are supported on runners that are secured to a single row of posts or uprights, and the inner ends are supported on or in a structure or building wall. A single-pole scaffold usually rests on one upright and is attached to the building for support.

Double-pole scaffolds consist of a platform that rests on cross beams (bearers) supported by ledgers and a double row of uprights that are independent of support (except ties, guys, or braces) from any structure. Double-pole scaffolds are supported independently of the structure.

The following requirements apply to single- and double-pole scaffolds:

- Runners and bearers must be installed on edge.

- Diagonal bracing must be installed in both directions across the entire outside face.

- Runners must extend over at least two poles and must be supported by bearing blocks securely attached to the poles.

- Braces, bearers, and runners must not be spliced between poles.

- Double-pole scaffolds must have cross-braces placed at the inner and outer sets of the poles.

In addition, there are very specific rules for the construction and support of wood-pole scaffolds. Although the residential construction industry often uses job-built or makeshift scaffolds, they rarely meet the safety requirements for pole scaffolds. Consult your competent person about the proper way to build pole scaffolds. Remember that all scaffolds must be able to withstand four times the intended load. Following is a list of common types of scaffolds found on residential construction sites and the requirements for each scaffold type.

Fabricated (tubular-welded) frame

This type of scaffold consists of platforms supported by fabricated end frames with integral posts, horizontal bearers, and intermediate members. The following requirements apply to fabricated frame scaffolds:

- When moving platforms to the next level, do not remove the existing platform until the new end frames are set in place and braced.

- Vertical members of frames must be secured laterally.

- Secure cross bracing so that the frame is automatically square and aligned.

- Frames and panels must be joined by couplings, stacking pins, or other means to keep the frame secure.

- Use locking pins or similar protection to prevent frame uplift if needed.

Horse

Horse scaffolds consist of a platform supported by construction horses (saw horses). The following requirements apply to horse scaffolds:

- Horse scaffolds should not have more than two levels or be over 10 ft. high.

- The second level must be placed directly on top of the first and be cross-braced with the legs nailed down or secured to prevent the scaffold from moving.

Carpenter's bracket

This consists of a platform supported by brackets that are attached to a building or to structural walls. The following requirement applies to carpenter's bracket scaffolds:

- A bolt extending through the wall can be used as the structural connection.

Top plate bracket

This consists of a platform supported by brackets that hook over or are attached to the top of a wall. This type of scaffold is similar to the carpenter's bracket and form scaffolds and is used for setting trusses in residential construction.

Form

This scaffold consists of a platform supported by brackets attached to form work. The following requirement applies to form scaffolds:

- Must be attached to the structure by nails, metal stud attachments, welding, or hooking over a secured structural support member with the form wales bolted or secured to the structure, and the snap ties or tie bolts going through the form.

Roof bracket

This scaffold consists of a platform that rests on angular supports. If you are using roof bracket scaffolds as roof-fall protection, they must be installed and used as "slide guards." The following requirements apply to roof bracket scaffolds:

- The roof bracket must fit the pitch of the roof it is used for, and the platform must be level.
- Brackets must be secured to the roof according to manufacturer's directions.

Pump jack

This consists of a platform supported by vertical poles and movable support brackets. The following requirements apply to pump jack scaffolds:

- Brackets, braces, and accessories must be made from metal plates and angles.
- Each bracket must have two gripping mechanisms to prevent slippage.
- Poles must be attached to the structure with rigid, triangle braces at the top and bottom and at any other locations necessary to keep the scaffold secure.
- When a pump jack must pass over a brace that has already been installed, another brace must be installed about 4 ft. above the brace that has to be passed. After the pump jack has passed the original brace location, the brace can be reinstalled.
- Workbenches may be used for the top rail of a guardrail system as long as they are the right height and strength.
- Workbenches should not be used as work platforms.
- Wood poles must be straight-grained and free of shakes; large, loose, or dead knots; and other defects that could weaken the pole.
- Wood poles that are built of two continuous lengths must be joined together with the seams parallel.
- When 2×4s are spliced together to make a pole, the mending plates must maintain the full strength of the lumber.

Ladder jack

This consists of a platform that rests on brackets attached to two ladders.

- Ladder jack platforms must not exceed 20 ft. in height and must not be bridged together.
- Job-made ladders must not be used to support a scaffold.
- Ladders must be properly manufactured and must be sturdy.
- The ladder jack must be assembled so that the weight bears on the siderails and ladder rungs.
- The bearing area must be at least 10 in. long on each rung.
- Ladders used to support ladder jacks must be equipped with devices to prevent slipping.
- A personal fall arrest system (PFAS) is required for work higher than 10 ft.

Window jack

This scaffold consists of a platform that rests on a bracket or a jack that projects through a window opening.

- The window jack bracket must be attached to the window opening and should be used only for working at that window opening.
- Window jacks must not be used to support planks between one window jack and another.

Crawling boards (chicken ladders)

This type consists of a plank with cleats that are spaced and secured to provide footing. This type of scaffold is used on sloped surfaces, such as roofs.

- Crawling boards must extend the full length of the roof from peak to eave.
- Boards must be secured to the roofs.
- If a 34-in. grab line is secured along the length of the crawling board, then a guardrail or PFAS is not needed.

Step, platform, and trestle ladder

These scaffold types consist of platforms that rest directly on the rungs of stepladders or trestle ladders.

- Platforms must not be placed higher than the next-to-last rung or step.
- Scaffolds must not be bridged together.
- All ladders used in conjunction with step, platform, or trestle ladder scaffolds must be properly manufactured and must support the scaffolds.
- Do not use job-built ladders to support scaffolds.
- All ladders used to support step, platform, and trestle scaffolds must be equipped with devices to prevent slipping.

Mobile

This is a powered or unpowered portable, caster or wheel-mounted supported scaffold.

- Cross-, horizontal, and diagonal braces must be secured to prevent collapse.
- Vertical members must be laterally secured so that the scaffold is automatically plumb and aligned.

- Casters must be locked to stop accidental movement, with caster stems and wheels secured.

- Platforms must not extend beyond the supports unless stability is maintained.

- Screw jacks or similar devices must be used to level the scaffold.

- When moving the scaffold, the base should be pushed on or near the bottom but not higher than 5 ft. above the floor.

- Scaffolds must be stabilized during movement to prevent tipping.

- Avoid moving scaffolds while workers are on them. If this cannot be avoided, make sure there are no pits, holes, or obstructions on the floor. The angle of the floor must be within 3 degrees of level.

- The height-to-width base ratio is 2:1 or less, unless the scaffold meets the requirements of Appendix A of OSHA's scaffold regulation.

- When moving a mobile scaffold, workers are not permitted beyond the end supports, and workers must be told the scaffold is moving before it can be moved.

Training requirements

The competent person must train all employees who will be working with scaffolds. The safe work practices training must include the following:

- proper techniques for erecting, dismantling, moving, operating, repairing, and maintaining scaffolds
- electrical hazards
- fall protection systems
- falling object protection systems
- proper use of scaffolds
- materials handling on scaffolds
- load-carrying capacities
- design criteria

Employees must be retrained if they demonstrate a deficiency in any of these areas or when new hazards appear on the job, such as changes in the use of a scaffold fall protection system or a falling object protection system.

Safe work practices

All employees must adhere to the following safe work practices.

General guidelines

- Wear a hard hat anytime you are working on or near scaffolds.
- Build all scaffolds according to the competent person and manufacturer's directions.
- Ensure that each scaffold is strong enough to support the platform and at least four times the expected load.
- Build all working-level scaffold platforms at least 18 in. wide.

OSHA REGULATIONS AND SAFE WORK PRACTICES

- Deck the platform with no more than a 1 in. space between the decking/platform units and the upright supports. If there is not enough space to fully plank/deck, then you must plank/deck as much as possible.

- Deck as much as necessary to protect yourself when using the platform as a walkway, or for employees who will be erecting or dismantling the scaffold.

  - *Exception:* The decking/platforms for ladder-jack, pump-jack, top-plate, and roof bracket scaffolds can be as narrow as 12 in. wide.

- Make the decking as wide as possible if there is not enough space to build the minimum platform size.

- Keep the front edge of the platform within 14 in. of the face of the work. If this is not possible, you must use guardrails or PFASs to keep from falling to the inside of the work area.

  - *Exception:* The distance between the edge of the platform and the face of the work can be 18 in. for plastering or lathing.

- Cleat or attach platforms to the scaffold or make the planking extend at least 6 in. past the supports. If a platform is shorter than 10 ft., the platforms must not extend past the supports by more than 12 in. unless there is support for the cantilevered section. Platforms longer than 10 ft. must not extend past the supports by more than 18 in. unless there is support for the cantilevered sections. If you can't access those cantilevered sections, you don't have to support them.

- Build longer platforms with the abutting ends of the plank/deck resting on separate supports, or secure them by some other means.

- Overlap the ends of planks/decking by 12 in. on the supports, or nail or secure the ends together by some other means.

- Do not paint the top or bottom of work platforms with anything that will hide defects. Paint only the sides for identification.

- The competent person must decide if it is safe to intermix scaffold parts.

Access

- Use portable, hook-on, or attachable ladders to access the scaffold when the platform is more than 2 ft. above or below the access point. You can also have direct access from another scaffold or the actual structure as long as it is not more than 14 in. away.

- Don't use cross braces to climb on or off scaffolds.

- Place portable, hook-on, and attachable ladders securely to prevent the scaffold from tipping. Make sure that the bottom rung is no more than 24 in. above the ground or floor.

- Use the proper ladder to access the scaffold you are using. Rungs must be at least 12 in. long with maximum spacing of 12 in. between rungs.

Fall protection

- Use a PFAS when working on ladder jacks that are more than 10 ft. above the ground.
- Use a guardrail, PFAS, or grab rope alongside a crawling board/chicken ladder.
- Use fall protection (guardrails) on all scaffolds that are more than 10 ft. high.
- Use guardrails along all open sides and ends and build to the following requirements:
  - Top rails between 39 and 45 in. high must be installed.
  - Midrails must be installed halfway between the platform and the top rail. If using mesh or panels, install them from the top to bottom of the guardrail.
  - Guardrails must withstand 200 lb. of downward force and must not be made of steel or plastic banding.
  - Rail ends must not hang over the edge of scaffolds.
  - Midrails and mesh must withstand at least 150 lb. of downward force.
  - Manila or plastic rope can be used as a guardrail only if it is inspected by the competent person and meets the criteria for guardrails.
  - Cross bracing can be used in place of top rails or midrails (but not both at the same time) if the cross is between 20 and 30 in. above the platform for the midrail or 38 to 48 in. above the platform for the top rail.
  - Surface the guardrails to prevent puncture wounds or lacerations.

Falling object protection

- Wear a hard hat anytime you are working on or near scaffolds.
- Keep objects away from the edge of the scaffolds.
- Build toe boards to a force of at least 50 lb. with a minimum 4 in. in height. If material is taller than the toe boards, netting or other control measures will need to be put into place.

Additional rules for scaffold use

- Do not use any part of a scaffold that is damaged or weakened.
- Do not work on scaffolds if you feel weak, sick, or dizzy.
- Do not work on any part of the scaffold other than the work platform.
- Do not alter the scaffold.
- Do not move a scaffold horizontally while employees are on it unless it is a mobile scaffold and the proper procedures are followed.
- Do not perform heat-producing activities such as welding without taking precautions to protect scaffold components.
- Do not work on scaffolds that are covered with snow, ice, or other slippery matter.

- Do not work on or from scaffolds during storms or high winds unless the competent person has determined that it is safe to do so.

- Do not allow debris to accumulate on platforms.

- Do not overload scaffold platforms.

- Do not use makeshift devices, such as boxes and barrels, on top of scaffold platforms to increase the working level height.

- Do not erect, use, alter, or move scaffolds within 10 ft. of overhead power lines.

- Do not use shore or lean-to scaffolds.

- Do not swing loads near or on scaffolds unless you use a tag line.

FALL PROTECTION

Falls are the leading cause of fatal injuries in the construction industry. According to the U.S. Department of Labor's Bureau of Labor Statistics (BLS), 1,186 fatalities occurred in the construction industry in 2005—394 of these fatal incidents resulted from falls. In addition to being the leading cause of fatal injuries, falls also represent a significant portion of serious injuries in the residential construction industry. Deaths and serious injuries from falls occur for many different reasons, including falling from a ladder or scaffolding, off a roof, or through an unprotected floor hole from a piece of machinery or vehicle. To limit fall hazards, you should

- select appropriate fall protection systems

- ensure that safety systems are properly constructed and installed

- implement and supervise safe work practices

- train all employees in the selection, use, and maintenance of fall protection systems

Employer responsibilities

Employers are required to protect employees from falls of 6 ft. or greater. On most interior fall hazards, conventional fall protection methods (guardrails, safety nets, PFASs, or covers) will provide effective protection. Conventional fall protection is effective in providing protection for

- many walking or working surfaces
- floor holes
- wall openings
- skylight openings
- HVAC openings
- unprotected sides and edges

Guardrails and covers are the most commonly chosen protection systems in residential construction. To ensure that protection is adequate, the guardrails and covers must be installed according to OSHA's requirements.

Written fall protection plan

Written policies regarding fall protection should be a part of your comprehensive safety manual. Your fall protection plan should outline the roles of supervisors and workers and identify how and where fall protection will be used on the jobsite. The plan should be specific to your operation and the fall hazards workers may encounter. It should be reviewed regularly and revised when necessary.

- setting and bracing of roof trusses/rafters
- installing floor sheathing and joists
- performing roof sheathing operations
- erecting exterior walls
- installing foundation walls
- conducting attic and roof work
- applying roofing materials

The safe work practices listed in this chapter address preventing falls to lower levels. You must conduct regular safety checks and be diligent in enforcing policies and procedures. Remember, you are responsible for immediately correcting any unsafe practices or conditions.

Employees must be trained in fall protection and be committed to following safe work practices, except when doing so would expose the employee to a greater hazard. If this is the case, the employee must communicate his or her concern to the competent person in charge of fall protection before proceeding.

All fall protection measures must be installed as soon as possible and must remain in place as long as the hazard exists. Employers must ensure that drywall installers do not remove any floor, window, or stair guardrails during their work. The employer must designate a competent person to monitor the safety of employees working on flat roofs. This person must be close enough to communicate orally with the employee and must not have any other duties that could distract him or her from their safety monitoring function.

Controlled access zones

When using any alternative safe work practice, employees must be protected through limited access to high-hazard locations. The competent person must define the area where the hazard exists as a controlled access zone (CAZ). The CAZ must be cordoned off by signs, wires, tapes, ropes, or chains.

- All access to the CAZ must be restricted to authorized entrants.
- All employees who are permitted in the CAZ must be listed or be visibly identifiable by the competent person before they enter the zone.
- All protective elements of the CAZ must be enforced before beginning work.

Roof truss/rafter erection

During the setting of roof trusses or erecting rafters, the time it takes to properly erect and dismantle guardrails, PFASs, or scaffolding could expose employees to a greater fall hazard than actually setting the trusses/rafters. Therefore, a fall protection plan should be implemented and alternative safe work practices should be used. Only properly trained employees should be allowed to install roof trusses or erect rafters.

A combination of the following methods should be used to protect employees while setting trusses or erecting rafters:

- Employees must be trained in the safe work practices for working on or in the truss.
- Employees should not have other duties while setting trusses or erecting rafters.
- Once truss or rafter installation begins, employees who are not involved in that activity must not stand or walk where they could be struck by falling objects.
- The first two trusses or rafters should be installed from scaffolding or ladders and secured in place.
- Ensure that the wall can support the weight of the ladder leaning on side walls or a scaffold that may be attached to a wall.
- After the first two trusses or rafters have been set, an employee can climb onto the interior wall top plate or previously stabilized trusses to brace or secure the peaks.
- Safe access to the truss/rafter must be provided.
- Employees should use the previously stabilized truss/rafter as a support while setting each additional truss/rafter.
- All trusses/rafters must be adequately braced before any employee can use the truss/rafter as a support.
- Employees can only move onto the next truss/rafter after it has been secured.
- Walking and working on the exterior wall top plate is not permitted.
- Employees positioned at the peak or in the webs of the trusses must work from a stable position either by sitting on a ridge seat or by positioning themselves in a previously stabilized truss or rafter.

Follow these additional safe work practices when working inside the trusses.

- Only trained employees may be allowed to work at the peak during roof truss or rafter installation.
- Employees may perform work inside truss areas such as installing bracing or making repairs.
- All trusses should be adequately braced before any employee can use the truss as a support.
- Safe access into the truss area must be provided.
- When practical, temporary flooring should be installed to create a safe, solid work platform if ongoing work will be performed in the attic area.
- Keep both hands free when climbing through the truss areas.
- In order to avoid laceration injuries, do not place hands near metal gusset plates of roof trusses.
- Employees who are securing and bracing trusses or detaching trusses from a crane should

- have no other duties
- work at the peaks or in the webs of trusses in a stable position by working in previously braced trusses
- not remain on the peak or in the truss any longer than necessary to safely complete the task

Interior scaffolding, such as sawhorse or trestle ladder, or exterior scaffolding, such as top plate, can be used as a safe alternative when setting trusses or erecting rafters. Various types and brands of interior and exterior scaffolding are commercially available. Be sure to follow the manufacturer's safety instructions for all scaffolding. Interior scaffolding can be installed along the interior wall below the area where the trusses/rafters will be set and work can be done from this established platform. For exterior scaffolding, walls that support the scaffold must be capable of supporting, without failure, their own weight plus four times the maximum intended load on the scaffolding.

Key elements of scaffolding safety include the following:

- Scaffolding that is 10 ft. or higher must be equipped with guardrails.
- Scaffolding planks must be secured so that they do not move.
- Interior scaffolding may consist of saw horse scaffolding, planks supported by ladders, or planks supported by top plate brackets.
- Scaffolding set up and take down must be conducted under the supervision of a competent person.

Another option for safe installation of trusses is to position the truss system on the ground. Sheath and brace the trusses for stability. The fully assembled roof system can be lifted and set in place with a forklift or crane. Remember to follow the manufacturer's instructions for safe forklift and crane operation.

Attic work

The following steps must be taken to protect employees when exposed to fall hazards in attics and on roofs while installing drywall, insulation, HVAC systems, electrical systems (including alarms, telephone lines, and cable TV), plumbing, and carpentry (the application of shingles, tile, and other roof covering is covered under roof materials application):

- Allow only trained employees to work in these areas.
- Use a proper size ladder to gain access to the attic area.
- If practical, establish a platform for employees to work from. Place approved plywood or metal planking that meets the standards of the American National Standards Institute (ANSI) in the attic to provide a walking/working area.
- Materials and other objects that could be impalement hazards must be kept out of the area below or properly guarded.
- Materials and equipment for the work must be located in close proximity to the employee.
- Use caution and avoid touching metal gusset plates of the roof trusses.

- Use extra caution when working above foyers or stairways as the potential fall distance is greater. Avoid working in these areas when possible.

- When attic or roof work is in progress, employees should not stand or walk below or adjacent to any openings in the ceiling where they could be struck by falling objects.

- Operations must be suspended when adverse weather, such as high winds, rain, snow, or sleet, creates a hazardous condition.

Building and erecting exterior walls

During the construction and erection of exterior walls, the following steps must be taken:

- Employees must attend training before erecting exterior walls.

- A painted line 6ft. from the floor deck edge must be clearly marked before any wall erection activities to warn of the approaching unprotected edge.

- Materials for operations must be staged to minimize fall hazards.

- Build exterior walls with as much cutting of materials and other preparation as far away from the edge of the deck as possible.

- Do not walk backward toward the edge of the deck.

- Walk and work at least 6 ft. from the deck's edge whenever possible.

- Stop work if weather conditions create an unsafe condition.

- Install guardrails at window/wall openings prior to the placement of the wall section or immediately following the placement of the wall when there is a potential for a fall 6 ft. or more to a lower level through the opening.

- If at all possible, use a forklift to place the wall in position.

Foundation formwork and block walls

The following steps must be taken to protect employees when exposed to fall hazards while working from the top surface of block or concrete foundation walls, and related formwork:

- Employees must attend training before working on the top of the foundation wall/formwork and only as necessary to complete the construction of the wall.

- Employees should consider working from a work platform, such as a scaffold attached to the concrete forms or a mobile scaffold.

- All formwork must be adequately supported before employees can be on top of the formwork.

- When hazardous conditions exist, such as high wind, rain, sleet, or snow, suspend operations on foundation/formwork until the hazardous condition no longer exists.

- Use a safe means to gain access to foundation walls. This access may include a ladder or a secured access plank.

- Consider using multiple access points to reduce the travel distance required on top of the foundation wall.

- Materials and equipment must be located in close proximity to the employee when on top of the foundation/formwork.

- Materials and other objects that could be an impalement hazard must be properly guarded or kept out of the area below.

Floor joists/trusses installation

During the installation of floor sheathing/joists, the following steps must be taken:

- Employees must attend training before installing floor joists or sheathing.

- Materials must be staged to allow for easy access.

- To minimize fall hazards where basement foundations are used, employees must work/walk on the ground outside the foundation wall, where possible, rather than walk/work on top of the wall.

- Do not walk on top of the foundation plating until it has been fastened to the foundation wall.

- Employees should work from ladders when setting joists on top of the I-beams spanning the basement rather than walking on top of the I-beams.

- First floor joists/trusses must be rolled into position and secured from the ground, ladders, or scaffolds.

- Each successive floor joist/truss must be rolled into place and secured from a platform created from a sheet of plywood laid over the previously secured floor joists or trusses.

- Except for the first row of sheathing, which must be installed from ladders or the ground, employees must work from the established deck.

- Employees not assisting in the leading edge construction while leading edges still exist (e.g., cutting the decking for the installers), must not go within 6 ft. of the leading edge under construction.

Installation of floor sheathing

- Only trained employees should be allowed to install floor sheathing.

- When installing the first row of sheathing/decking, employees should work from the ground, a ladder, or scaffold.

- After completing the first row of sheathing, employees should work from that established deck.

- Do not walk backward toward the edge of the deck.

- Protect any floor hole created by the sheathing operations with guardrails or covers, such as a fireplace hole or stair hole.

- Employees not assisting in the leading edge construction (e.g., cutting the decking for the installers) should not be permitted to work within 6 ft. (1.8 m) of the fall hazard.

Roof materials application

Roofers are likely to be exposed to significant fall hazards that can lead to serious injury or death. Special precautions should be taken for roofing work, which consists of the removal, repair, or installation of weatherproofing roofing materials, such as shingles, tiles, and tar paper.

Consider the following safe work practices during roofing operations.

- Employees installing shingles and other roofing material should utilize a PFAS during this operation.

- Read all manufacturer's instructions and warnings before using fall protection equipment.

- Anchor points used for personal fall arrest must be capable of supporting 5,000 lb. (2,273 kg) per employee and installed at a secure place on the roof and according to manufacturer's requirements.

- Ensure that the anchor point is located at a height that will not allow the employee to strike a lower level should a fall occur.

- Provide the safe ladder access to the roof. If a ladder is used for access, the ladder must extend 3 ft. above the roof line and be secured from movement.

- Load and store roofing materials on the deck in a manner that prevents the materials from sliding off the roof.

- A competent person must determine the proper method of storing material to avoid overloading the roof deck/trusses.

- Provide safe access to the roof level for stocking materials.

- Do not ride a material conveyor or crane load to gain access to the roof.

- Lower material from the roof deck to the ground in a safe manner to avoid injuries or property damage.

- Establish approved "drop zones" where walking or working is prohibited. An approved drop zone cannot be located above an entrance or exit to a home.

- Provide a spotter on the ground level to create an added layer of protection to prevent anyone from walking into the drop zone area.

Roof sheathing operations

Typically, employees install roof sheathing after all trusses/rafters and any permanent truss bracing is in place. Roof structures are unstable until some sheathing is installed. Conventional fall protection systems should not be used to protect employees until it is determined that the roofing system can be used as an anchor point.

All employees must prepare their footwear (clean mud or other slip hazards from shoes or boots) before they attempt to walk on the sheathing. Materials must be staged prior to work being performed to allow for the sheathing to be quickly installed and to minimize the time employees are exposed to a fall hazard.

The following steps must be implemented to protect employees from fall hazards while installing roof sheathing:

- Roof sheathing can only be installed by qualified employees.

- Employees should not work on the roof level if conditions are windy, snowy, icy, muddy, or otherwise hazardous.

- Sheathing operations should be suspended if wind creates an unsafe condition.

- Employees should wear shoes that provide maximum traction on the roof level.

- Once roof sheathing installation begins, employees not involved in that activity should not stand or walk below or adjacent to the roof opening or exterior walls in any area where they could be struck by falling objects.

- Employees should keep their shoes free of mud or snow while working on the roof level.

- Material used in the roof sheathing process should be located nearby to minimize employees' exposure to falls while obtaining material.

- The competent person must stop work as needed to allow passage through the area when a work stoppage would not create a greater hazard.

- Employees should stand in truss webs to install the bottom row of roof sheathing.

Using slide guards

After the bottom row of roof sheathing is installed, a slide guard extending the width of the roof must be securely attached to the roof per the following requirements:

- Slide guards must be constructed of 2 × 4 or 2 × 6 lumber capable of limiting an uncontrolled slide.

- Employees should install the slide guard while standing in truss webs and leaning over the sheathing.

- Additional rows of roof sheathing may be installed when employees are positioned on previously installed rows of sheathing. A slide guard can help employees retain their footing during successive sheathing operations.

- Additional slide guards must be securely attached to the roof at intervals not to exceed 13 ft. as successive rows of sheathing are installed.

- Roofs with pitches in excess of 9:12 must have slide guards installed at 4 ft. intervals.

- In wet weather (rain, snow, or sleet), roof sheathing operations must be suspended unless safe footing can be ensured.

- When strong winds (above 40 mph) are present, roof sheathing operations must be suspended unless windbreakers are erected.

Training requirements

You must provide training for all employees who might be exposed to fall hazards. The training should teach employees to recognize fall hazards and methods to minimize these hazards. Employees should only use the following alternative safe work practices after they have received fall hazard protection training.

Safe work practices

All employees must adhere to the following safe work practices.

Interior falls and guardrails

- Install guardrails or covers whenever there is a fall potential of 6 ft. or more.

- Make guardrail top rails 42 in. ± 3 in. above the walking/working level.

- Raise the top edge height of the top rail equal to the stilt height for an employee who is using stilts.

- Keep midrails halfway between the top edge of the guardrail system and the walking/working level (21 in. high).

- Build guardrail systems to withstand a force of at least 200 lb. (downward or outward thrust) along the top edge.

- Surface the guardrail to prevent injury or clothes snagging.

- Do not use steel banding and plastic banding as top or midrails.

- Erect toe boards along the edge of the walking/working surface.

- Build toe boards to a force of at least 50 lb. with a minimum 4 in. in height. If material is taller than the toe boards, netting or other control measures will need to be put into place.

- Build guardrails to the following specs:

  - *For wood railings.* Wood components must be minimum 1,500 lb.–ft./in. (2) fiber (stress grade) construction grade lumber; the posts must be at least 224 lumber spaced not more than 8 ft. apart on center; the top rail must be at least 224 lumber, the intermediate rail must be at least 126 lumber. All lumber dimensions are nominal sizes as provided by the American Softwood Lumber Standards (January 1970).

  - *For pipe railings.* Posts, top rails, and intermediate railings must be at least 1½ in. nominal diameter (schedule 40 pipe) with posts spaced not more than 8 ft. apart on center.

- Guard all unprotected holes with rails or covers.

- Color code or mark the word "hole" or "cover" on the cover.

Controlled access zones (CAZ)

- All access to the CAZ must be restricted to authorized entrants.

- All employees who are permitted in the CAZ must be listed or be visibly identifiable by the competent person before they enter the area.

- All protective elements of the CAZ must be enforced prior to beginning work.

Attic and roof work

- Materials and equipment must be kept in close proximity to the work area.

- Materials and other objects that could be impalement hazards must be kept out of the area below, or else such materials must be properly guarded.

- When attic or roof work is in progress, employees should not stand or walk below or adjacent to any openings in the ceiling where they could be struck by falling objects.
- Operations must be suspended when adverse weather (such as high winds, rain, snow, or sleet) creates a hazardous condition.

Erection of exterior walls

- Attend training before erecting exterior walls.
- Paint a line 6 ft. from the floor deck edge before any wall erection activities to warn of the approaching unprotected edge.
- Stage materials to minimize fall hazards.
- Perform cutting of materials and other preparation as far away from the edge of the deck as possible.

Foundation walls/formwork

- Attend training before you work on the top of the foundation wall/formwork and only as necessary to complete the construction of the wall.
- Ensure that all formwork is adequately supported before you get on top of the formwork.
- Operations must be suspended during inclement weather such as high winds, rain, snow, or sleet.
- Materials and equipment for the work must be kept in close proximity to employees who are on the top of the foundation/formwork.
- Materials and other objects that could be an impalement hazard must be properly guarded or kept out of the area below.

Floor joists/trusses and sheathing

- Attend training before you install floor joists or sheathing.
- Stage materials to allow for easy access.
- Roll first floor joists/trusses into position and secure them from ground, ladders, or sawhorse scaffolds.
- Roll each successive floor joist/truss into place and secure them from a platform created from a sheet of plywood laid over the previously secured floor joists or trusses.
  - *Exception:* The first row of sheathing work from the established deck, which must be installed from ladders or the ground.
- If not assisting in the leading-edge construction while leading edges still exist (e.g., cutting the decking for the installers), do not go within 6 ft. of the leading edge under construction.

Roof materials application

- Employees should utilize a PFAS when installing shingles and other roofing material.
- Employees should read all manufacturer's instructions and warnings before using fall protection equipment.

- Anchor points used for personal fall arrest must be capable of supporting 5,000 lb.(2,273 kg) per employee and installed at a secure place on the roof and according to manufacturer's requirements.

- Ensure that the anchor point is located at a height that will not allow the employee to strike a lower level should a fall occur.

- Provide safe ladder access to the roof. If a ladder is used for access, the ladder must extend 3 ft. above the roof line and be secured from movement.

- Load and store roofing materials on the deck in a manner that prevents the materials from sliding off the roof.

- A competent person must determine the proper method of storing material to avoid overloading the roof deck/trusses.

- Provide safe access to the roof level for stocking materials.

- Do not ride a material conveyor or crane load to gain access to the roof.

- Lower material from the roof deck to the ground in a safe manner to avoid injuries or property damage.

- Establish approved "drop zones" where walking or working is prohibited. An approved drop zone cannot be located above an entrance or exit to a home.

- Provide a spotter on the ground level to create an added protection factor to prevent anyone from walking into the drop zone area.

Roof sheathing operations

- Do not install roof sheathing unless you are qualified to do so.

- Employees who are not involved in the roof sheathing installation should not stand or walk below or adjacent to the roof opening or exterior walls or in any area where they could be struck by falling objects.

- The competent person must define the limits of this area before sheathing begins.

- The competent person must stop work as needed to allow passage through such areas when this work stoppage would not create a greater hazard.

- The bottom row of roof sheathing may be installed by workers standing in truss webs.

When using slide guards:

- After the bottom row of roof sheathing is installed, a slide guard extending the width of the roof must be securely attached to the roof.

- Slide guards must be constructed of 2 × 4 or 2 × 6 lumber capable of limiting an uncontrolled slide.

- Install the slide guard while standing in truss webs and leaning over the sheathing.

- Additional rows of roof sheathing may be installed when you are positioned on previously installed rows of sheathing. A slide guard can help you retain your footing during successive sheathing operations.

- Additional slide guards must be securely attached to the roof at intervals not to exceed 13 ft. as successive rows of sheathing are installed.

- Roofs with pitches in excess of 9:12 must have slide guards installed at 4 ft. intervals.

- In wet weather (rain, snow, or sleet), roof sheathing operations must be suspended unless safe footing can be ensured.

- Suspend roof-sheathing operations during winds above 40 mph unless windbreakers are erected.

Roof truss/rafter erection

- Employees must be trained in the safe work practices for working on or in the truss.

- Employees should not have other duties while setting trusses or erecting rafters.

- Once truss or rafter installation begins, employees who are not involved in that activity must not stand or walk where they could be struck by falling objects.

- The first two trusses or rafters should be installed from scaffolding or ladders and secured in place.

- Ensure that the wall can support the weight of the ladder leaning on side walls or a scaffold that may be attached to a wall.

- After the first two trusses or rafters have been set, an employee can climb onto the interior wall top plate or previously stabilized trusses to brace or secure the peaks.

- Safe access to the truss/rafter must be provided.

- Employees should use the previously stabilized truss/rafter as a support, while setting each additional truss/rafter.

- All trusses/rafters must be adequately braced before any employee can use the truss/rafter as a support.

- Employees can only move onto the next truss/rafter after it has been secured.

- Walking and working on the exterior wall top plate is not permitted.

- Employees positioned at the peak or in the webs of the trusses must work from a stable position either by sitting on a ridge seat or by positioning themselves in a previously stabilized truss or rafter.

- Employees should not stay on the peak/ridge any longer than necessary to safely complete the task.

Follow these additional safe work practices when working inside the trusses.

- Only trained employees must be allowed to work at the peak during roof truss or rafter installation.

- Employees may perform work inside truss areas such as installing bracing or making repairs.

- All trusses should be adequately braced before any employee can use the truss as a support.

OSHA REGULATIONS AND SAFE WORK PRACTICES

- Safe access into the truss area must be provided.

- When practical, temporary flooring should be installed to create a safe, solid work platform if ongoing work will be performed in the attic area.

- Keep both hands free when climbing through the truss areas.

- In order to avoid laceration injuries, do not place hands near metal gusset plates of roof trusses.

- Employees who are securing and bracing trusses or detaching trusses from a crane should
  - have no other duties
  - work at the peaks or in the webs of trusses in a stable position by working in previously braced trusses
  - not remain on the peak or in the truss any longer than necessary to safely complete the task

Interior scaffolding, such as sawhorse or trestle ladder, or exterior scaffolding, such as top plate, can be used as a safe alternative when setting trusses or erecting rafters. Various types and brands of interior and exterior scaffolding are commercially available. Be sure to follow the manufacturer's safety instructions for all scaffolding. Interior scaffolding can be installed along the interior wall below the area where the trusses/rafters will be set and work can be done from this established platform. For exterior scaffolding, walls that support the scaffold must be capable of supporting, without failure, their own weight plus four times the maximum intended load on the scaffolding.

Key elements of scaffolding safety include the following:

- Scaffolding that is 10 ft. or higher must be equipped with guardrails.

- Scaffolding planks must be secured so that they do not move.

- Interior scaffolding may consist of sawhorse scaffolding, planks supported by ladders, or planks supported by top plate brackets.

- Scaffolding set up and take down must be conducted under the supervision of a competent person.

Another option for safe installation of trusses is to position the truss system on the ground. Sheath and brace the trusses for stability. The fully assembled roof system can be lifted and set in place with a forklift or crane. Remember to follow the manufacturer's instructions for safe forklift and crane operation.

CRANES

The use of cranes in the residential construction industry is not as frequent as in commercial construction. Most builders only use cranes for specific jobs such as setting roof trusses. Many vendors use cranes to unload materials on the jobsite. Whatever the reason for their use, it is important to remember that crane hazards can lead to electrocution, crushing, and structure or crane collapse. Even small cranes can be hazardous to employees if the proper safety requirements are not followed.

Employer responsibilities

Employers must follow the manufacturer's specifications and limitations for crane operation. Attachments used for cranes must not exceed the recommended capacity, rating, or scope. Before cranes can be used, all rated load capacities, recommended operating speeds, special hazard warnings, or instructions must be posted on the crane and visible to the operator while in the operating station. Hand signals used for crane operation must meet the ANSI standard for the crane type used, and an illustration of the signals must be posted at the jobsite.

A competent person must determine that the crane is safe prior to and during each use. OSHA has specific inspection requirements for cranes that must be observed prior to each use and throughout the life of the crane.

Crane structure requirements

All exposed belts; gears; shafts; pulleys; sprockets; fly wheels; chains; or other reciprocating, rotating, or other moving parts must be properly guarded to prevent employee contact. In addition, the swing radius of the crane superstructure needs to be adequately barricaded to prevent employees from being crushed or struck by the swing. The window glass in the cab of the crane must be safety glass, or equivalent, and must be free of cracks or breaks that could distort the operator's view and interfere with the safe operation of the crane. All cranes must have at least one 5-BC rated fire extinguisher on board at all times.

Overhead power lines and crane transport

It is critical that a safe distance be maintained between the crane/boom and overhead power lines. Unless power lines are de-energized or guarded in some manner, the operator must maintain a 10-ft. distance between the crane and the line. Distance requirements increase as the voltage increases. An observer, or *spotter*, must be used if there is a chance that the crane operator cannot accurately judge the distance between the crane/boom and the overhead power lines.

Training requirements

You must designate a competent person to inspect all machinery and equipment prior to its use on the site. The equipment must be monitored during use to ensure that it remains in safe operating condition.

Safe work practices

All employees must adhere to the following safe work practices.

- Do not walk under crane loads.
- Follow the crane operator's and signaler's instructions.
- Stay clear of the crane's superstructure.

- Stay a safe distance away from cranes in operation if your work does not require you to be in the area.
- Hand signals used for crane operations must meet the ANSI standard for crane type used and an illustration of the signals must be posted at the jobsite.

MOTOR VEHICLES

The use of motor vehicles and mechanized equipment is essential in the residential construction industry. Whether employees are driving company vehicles or using their own vehicles in the course of work—safe maintenance of motor vehicles and equipment is critical to protecting employees on the job. Employers and employees must work together to keep this equipment in proper working condition.

Employer responsibilities

It is your responsibility to ensure that all company-owned vehicles and mechanized equipment are equipped with safety systems. Mechanized equipment should be regularly inspected, and safety hazards should be repaired immediately. The following vehicle components should be inspected prior to use:

- brake systems (service brakes, trailer brake connections, hand brakes, and emergency stopping brakes)
- tires
- horn
- steering mechanism
- lights
- coupling devices
- seat belts (seat belts are not required for equipment used for stand-up operation)
- operating controls
- other safety systems

Training requirements

Although there are no specific OSHA requirements for motor vehicles and mechanized equipment, employers should observe the general OSHA training requirement that states: "Each employer shall instruct each employee in the recognition and avoidance of unsafe conditions and the regulations applicable to his work environment to control or eliminate any hazards or other exposure to illness or injury."

Safe work practices

All employees must adhere to the following safe work practices.

- Ensure that all off-road equipment used on site is equipped with rollover protection.
- Use a backup alarm or guide when operating in bidirectional machines such as rollers, compactors, front-end loaders, bulldozers, and similar equipment.
- Use a horn that can be distinguished from the surrounding noise when operating motor vehicles.
- Maintain the horn in operating condition.
- Ensure that all vehicles have fully operational braking systems and brake lights.

- Ensure that all vehicles have working headlights and taillights.

- Use seat belts when transporting employees in motor and construction vehicles.

- Seats must be firmly secured with enough seating for all employees being transported.

- Inspect all vehicle brake systems, tires, horns, steering mechanisms, lights, couplings, and seat belts (not required for equipment used for stand-up operation) before use for broken or unsafe conditions.

EXCAVATION AND TRENCHING

Excavations and trenches pose a serious hazard on many residential construction sites. Therefore, employees who work in and around excavations and trenches must understand the importance of cave-in protection, and employers must take precautions to prevent a catastrophe. Contrary to popular belief, employees can't survive a trench collapse by simply holding their breath, nor can they hear the dirt before it collapses and outrun it.

Employer responsibilities

Notifying the state's one-call system is the most overlooked step in excavation work. You must notify the state prior to beginning excavation, so that underground utilities can be located and marked. If that is not possible, extreme caution must be used while excavation operations are underway. Any identified utilities are to be supported, protected, or removed while the excavation is open.

You must designate a qualified employee to be the competent person. The competent person determines whether it is safe for employees to enter an excavation or trench. The decision to work in excavations or trenches must be based on the evaluation of soil type, soil stability, and other factors on which the competent person has training and experience. The competent person must also select the appropriate sloping, shoring, or benching protective system to protect employees.

Additionally, you must

- ensure that all surface equipment or materials that could affect the stability of the excavation are removed or supported

- ensure that employees use stairs, ladders, ramps, or equivalent access tools in any excavation over 4 ft. deep and locate access tools within 25 ft. of employees to aid in their exit

- prevent employees from standing beneath the load of any lifting or earth-moving equipment

- prevent employees from working in excavations where there is accumulated water, unless adequate precautions have been taken such as special support or shield systems to prevent cave-ins, water removal methods to control the level, or use of a safety harness and lifeline

- protect employees from loose rock or soil that could be a hazard from falling or rolling from the face of the excavation, including scaling to remove loose material, installation of protective barricades at necessary intervals to stop/contain falling material, or other methods that provide equal protection
- ensure that spoils, piles, and equipment are kept at least 2 ft. away from the edge of excavations

Training requirements

The competent person will conduct daily inspections of excavations, the adjacent areas, and protective systems for hazardous conditions prior to the start of work and as needed throughout the shift.

Safe work practices

All employees must adhere to the following safe work practices.

Protective systems

- Protect all excavations over 5 ft. deep with shoring, trench boxes, or sloping of the earth according to the direction of the competent person.
- Protect excavations less than 5 ft. deep with shoring, trench boxes, or the sloping of the earth if the competent person determines that a cave-in is a possibility.

General guidelines

- Remove or support all surface obstructions as needed.
- Request utility companies or owners to determine the location of underground utility installations before excavating.
- Ensure that underground installations are protected, supported, or removed while the excavation is open.
- Proceed with caution and use detection equipment if the exact location of lines is unknown.
- Use stairs, ladders, and ramps every 25 ft. to exit the excavation.
- Wear warning vests or other high-visibility materials when near traffic.
- Do not stand beneath loads handled by lifting or digging equipment.
- Keep clear of vehicles loading or unloading.
- Use a warning system (barricades, hand or mechanical signals, or stop logs) if the equipment operator cannot see near the edge of an excavation.
- Do not work in excavations where water is standing or accumulating unless special supports are used to prevent cave-ins.
- Keep all spoils piles and equipment at least 2 ft. away from the edge of the excavation.
- Use support systems such as shoring, bracing, or underpinning to protect all other structures from collapse.
- Remove loose soil or use barricades for protection inside excavations.
- Inspect excavations, adjacent areas, and protective systems for

- evidence of possible cave-ins
- indications of failure of protective systems
- hazardous atmospheres
- other hazardous conditions at the beginning of each workday or if the stability of the excavation could change

- Install walkways over excavations that must be crossed. A guardrail system must be in place if the walkway is 6 ft., or more, above the lower level.
- Barricade wells, pits, or shafts.

STAIRWAYS AND LADDERS

The use of stairways and ladders is a vital part of residential construction; however, the misuse of these systems can result in serious injuries. Safe work practices must begin when the ladder is first placed on the ground and continue as employees climb up and down the ladder. The condition in which ladders are maintained, the steps are surfaced, and the way employees handle them are critical parts of a stair and ladder safety program. Instituting safe work practices can help you limit fall distances and reduce risk.

Employer responsibilities

Ramps, stairs, or steps must be available when employees have to cross an elevation break of 19 in. or more to access other levels.

Stairs

Guardrails are required on all open stairways and handrails must be provided on all stairways four risers high or more. Temporary stairs must be built to OSHA specifications.

Ladders

Ladders must be able to withstand the intended loads imposed on them. Employers should purchase ladders from reputable manufacturers and avoid building job-made ladders, which seldom meet OSHA's rigorous requirements.

Ladders must be inspected periodically by a competent person to determine if there are visible defects. Broken or damaged ladders must be marked as damaged and removed from the jobsite.

Training requirements

Employees must be trained to recognize ladder and stairway hazards and provide procedures for minimizing these hazards.

Safe work practices

All employees must adhere to the following safe work practices.

General requirements for temporary stairs

- Use a stairway, ramp, or ladder whenever you must step 19 in. or more to gain access to another surface.
- Build temporary stairs between 30 and 50 degrees from horizontal.
- Make riser height and tread depth the same for each flight of stairs. Don't vary by more than ¼ in.
- Provide a platform when doors open directly onto a stairway, and make sure

the width of the platform is not reduced to less than 20 in.

- Keep hazardous projections, such as protruding nails, splinters, etc., out of the stairs.

- Fix slippery conditions before the stairs are used.

- Place wood or other materials in unfilled pan stairs to keep the surface even.

- Build treads of stairs with wood or other solid material and install them the full width and depth of the stair.

Stair rails and handrails

- Install a 36-in.-high stair/handrail on unprotected sides of stairs with more than 3 risers or more than 30 in. high. Measure the 36 in. from the tread surface up to the top of the rail.

- Install midrails halfway between the top edge of the stair rail system and the stairway steps.

- Use a handrail offset on winding/spiral stairs to keep employees from walking where the treads are less than 6 in. wide.

- Build stair/handrail systems to withstand 200 lb. of force in any downward or outward direction at any point along the top edge.

- Install handrails to act as a handhold between 30 and 37 in. on all stair systems even if there is a wall.

- Surface stair/handrails to prevent injury and clothes snagging.

- Build stair handrails in a manner that does not create a projection hazard beyond the edge of the rail.

- Build temporary stair/handrails about 3 in. away from where the drywall will be installed, so that the rails will not have to be taken down in order to install the drywall.

- Use guardrail system to protect sides and edges of stairway landings.

Ladders

The following safety requirements apply to all ladders, including job-made ladders:

- Use ladders only on stable and level surfaces unless they are secured to prevent movement.

- Keep the area around the top and bottom of ladders clear of debris.

- Place ladders on nonslip surfaces or use ladders that have slip-resistant feet.

- Inspect ladders before each use and after any activity that could have caused damage.

- Remove ladders with structural defects from service.

- Maintain ladders free of oil, grease, and other slipping hazards.

- Follow the manufacturer's rated capacity for ladder use.

- Use ladders only for their designated purpose.

- Do not move, shift, or extend ladders that are occupied.

- Do not work outside of the foot print of the ladder. Never let your belt buckle pass beyond either ladder siderail.

- Place ladders at an angle of 4:1 for every 4 ft. in height the ladder should be placed 1 ft. out.

- Extend ladders at least 3 ft. beyond the surface you are accessing so that you have a handhold when getting on and off.

- Do not use a stepladder as an extension ladder.

- Do not tie ladders together to create a longer ladder.

- Do not put ladders on scaffolds or other unstable platforms.

- Support both ladder rails equally.

- Keep ladders from coming within 10 ft. of energized power lines.

- Do not use the top two steps of a step ladder as a seat or a ladder.

- Do not climb the cross bracing on the back of stepladders.

- Use the three-point contact system when working on ladders. Face the ladder when climbing up or down, and use at least one hand to hold onto the ladder.

- Do not carry objects or loads that could cause you to fall.

- Be sure that ladder rungs and steps are parallel, level, and uniformly spaced.

- Be sure that rungs or steps of portable ladders are between 10 and 14 in. apart.

- Be sure that siderails for all portable ladders are at least 12 in. apart.

- Always use and lock the metal spreader or device that holds the front and back sections of a step ladder apart.

- Surface ladder components to prevent punctures, lacerations, or clothes snagging.

- Coat wood ladders with clear covering only. Identification or warning labels can be placed on one face of a siderail.

- Do not build or use job-made ladders unless they meet the exact ANSI and OSHA specifications.

THE HAZARD COMMUNICATION STANDARD (HAZCOM)

Your safety program must also include a written hazard communication program. The hazard communication standard has implications for every contractor and builder in the construction industry. Haz-Com applies to all chemicals that will be used on the jobsite that pose a danger to employees who will be working with or around hazardous chemicals. These chemicals can range from paint to concrete to wood dust.

You are not responsible for determining if a chemical is hazardous. Hazard determination is the responsibility of the product manufacturer, distributor, or importer. However, the HazCom standard does require employers to develop and implement a written program and a system for maintaining material safety data sheets (MSDS) and product labels and to provide training to all employees who could be exposed to a hazardous substance.

OSHA REGULATIONS AND SAFE WORK PRACTICES

Generally, most hazardous substances found on construction sites will fall into one of the following categories:

- flammables and combustibles
- compressed gases
- toxins (e.g., systemic poisons, dusts, fumes, corrosives, irritants)

Written program

You must develop and implement a written HazCom program and maintain the program at each jobsite. The program must be available to employees and/or their designated representatives upon request. Although it does not have to be lengthy or complicated, it must detail how the requirements for labels and other forms of warnings, MSDSs, and employee information and training will be met. It must also include a list of the hazardous chemicals used on the jobsite and explain how you will inform employees of chemical hazards related to nonroutine tasks. The written program must reflect what your company is doing to meet the HazCom requirements, by indicating who is responsible for various aspects of the program and where written materials will be made available to employees.

In a multiemployer situation, the written program must also detail

- how you will provide the other on-site employer(s) access to MSDSs
- how other employer(s) can protect their employees from hazards during normal operations and emergencies
- how you will inform the other employer(s) of your labeling system

You should keep the program in a looseleaf binder for ease in updating. The CD that accompanies this book includes a written hazard communication program that you can adapt for your company.

List of hazardous chemicals

HazCom requires that you list each hazardous chemical that you use. This list will eventually serve as an inventory of every chemical for which an MSDS must be maintained. The best way to prepare the list is to survey the workplace to identify hazardous chemicals in containers, read the supplier-provided labels for hazard information, and review purchasing records. Chemicals that are generated by work operations, such as welding fumes, dusts, and exhaust fumes must also be listed. Once you have compiled the list, you must obtain an MSDS for each chemical on the list.

Material safety data sheets

You are responsible for obtaining an MSDS for every hazardous chemical used on your jobsites. If you do not receive an MSDS with a shipment, it's your responsibility to request it. Make your request in writing as evidence of your good-faith effort to comply should OSHA inspect your company. In addition, you must request information about hazardous substances and chemicals from other on-site contractors when working in a multiemployer situation. You must designate someone to obtain and maintain an MSDS for every hazardous chemical.

Retail outlets such as hardware stores or lumberyards are only required to supply MSDSs to customers with commercial accounts. For all other customers, the retailer must supply the address and telephone number of the manufacturer from which the MSDS can be obtained. You should get into the habit of automatically requesting the MSDS with your purchase if you have a commercial account. If you do not have a commercial account, be sure to request the manufacturer's address and telephone number.

MSDSs must be accessible to all employees when they are on the jobsite and during their shift. Some employers keep MSDSs in a binder in a central location (e.g., in the pick-up truck on construction sites). As long as employees can get the information when needed, any approach may be used. You should not allow an employee to use any chemical for which you have not received an MSDS.

Manufacturers, importers, and distributors periodically review and revise MSDSs. It is your responsibility to ensure that you have the most current version. Suppliers will provide the revised MSDS by mail or with your next shipment. You should review the updated version of the MSDS to determine if additional training is necessary.

Product labels

The main purpose of the label is to provide the employee with an immediate source of information. Labels must be clearly written or typed in English and be displayed in visible location on the container. If you have a multilingual workforce, you may choose to translate labels to their language(s). If the label includes signs and symbols that you do not fully understand, contact the supplier for assistance. The primary information to be displayed on an OSHA-required label is the identity of the material and the appropriate hazard warning. The identity used by the supplier may be a common or trade name (Super Lube) or as a chemical name (Sodium Phosphate). The hazard warning is a brief statement of the hazardous effects of the chemical such as flammable or causes lung cancer. Labels can contain information regarding precautionary measures (do not use near open flame), but this is not a requirement. Smaller portable containers that have been filled with a chemical dispensed from an original, labeled container must also be properly labeled with the identity and appropriate hazard warning.

Employee training

Once you have established an MSDS and labeling system, you will need to train your employees on how to use the system. Remember, you must train all employees who may be exposed to hazardous chemicals when working. This includes employees who work directly with a hazardous substance and may be exposed through any route of entry (inhalation, skin contact, ingestion, etc.), and those who have the potential to be accidentally exposed.

The following topics must be covered prior to an employee being exposed to a

OSHA REGULATIONS AND SAFE WORK PRACTICES

hazardous substance (including when a new hazardous substance is brought to the worksite):

- methods and observations that may be used to detect the presence of hazardous chemicals (e.g., visual appearance and/or odor)
- physical and health hazards of chemicals on the jobsite
- hazard protection methods
- work practices
- emergency procedures
- personal protective equipment
- details of your HazCom program
- labeling system
- MSDS location
- how to use the hazard information

After the training has been completed, be sure to document the topic, the attendees, and the date of training.

Safe work practices

To ensure HazCom compliance you must do the following:

- Obtain a copy of the Hazard Communication Standard.
- Read and understand the requirements.
- Assign responsibility for tasks (maintaining MSDSs, maintaining labels, and conducting training).
- Prepare an inventory of chemicals.
- Ensure containers are labeled.
- Obtain MSDSs for each chemical.
- Prepare a written program.
- Make MSDSs available to workers.
- Conduct employee training.
- Establish procedures to maintain current program.
- Establish procedures to evaluate the effectiveness of your written program.

Safety and Health Training

This chapter discusses the need for safety training. It provides an outline for new hire orientation and stresses the importance of ongoing training.

Successful safety program administrators will attest to the fact that good training is one of the most valuable investments a company can make. It is important to stress not just training, but good training. There are several forms of training. It is important for employers to develop a plan for continuous training to help keep employees' skills sharp and their minds focused on safety.

OSHA REGULATIONS REQUIRING TRAINING FOR COMPLIANCE

The following training requirements are a compilation of the actual OSHA regulations that apply to the residential construction industry. The requirements are quoted as they appear in *OSHA's Construction Safety and Health Standards* (29 CFR 1926). For further information you should review the complete standard, available through BuilderBooks.com (www.builderbooks.com or 1-800-223-2665).

For ease of analysis, the requirements are divided into three major categories:

1. stated

2. implied (OSHA states the need for the services of a competent or qualified person)

3. adopted (OSHA references an ANSI, NEC, or similar consensus standard)

STATED REQUIREMENTS

Safety Training Education (1926.21). The employer shall instruct each employee in the recognition and avoidance of unsafe conditions and the regulations applicable to his [or her] work environment to control or eliminate any hazards or other exposure to illness or injury.

Employees [who] handle or use poisons, caustics, and other harmful substances shall be instructed regarding their safe handling and use, and be made aware of the potential hazards, personal hygiene, and personal protective measures required.

In jobsite areas where harmful plants or animals are present, employees who may be exposed shall be instructed regarding the potential hazards and how to avoid injury, and the first aid procedures to be used in the event of injury.

Employees [who] handle or use flammable liquids, gases, or toxic materials shall be instructed in the safe handling and use of these materials and made aware of the specific requirements contained in Subparts D, F, and other applicable subparts of this part.

Medical Services and First Aid (1926.50). In the absence of an infirmary, clinic, hospital, or physician that is reasonably accessible in terms of time and distance to the work site that is available for the treatment of injured employees, a person who has a valid certificate in first aid training from the U.S. Bureau of Mines, the American Red Cross, or equivalent training that can be verified by documentary evidence, shall be available at the work site to render first aid.

Asbestos (1926.1101). The employer shall institute a training program for all employees exposed to airborne concentrations of asbestos, in excess of the action level and/or excursion limit and shall ensure their participation in the program.

Hazard Communication Construction (1926.59). Employee training should, at least, include the following components:

- Methods and observations that may be used to detect the presence or release of a hazardous chemical in the work area
- Physical and health hazards of the chemicals in the work area
- Measures employees can take to protect themselves from these hazards
- Details of the hazard communication program to be developed by the employer

Lead in Construction (1926.62). The employer shall communicate information concerning lead hazards according to the requirements of OSHA's Hazard Communication Standard for the construction industry, 29 CFR 1926.59, including but not limited to the requirements concerning warning signs and labels, material safety data sheets (MSDS), and employee information and training.

Respiratory Protection 1926.103 (1910.134). Employees required to use other types of respiratory protective equipment shall be instructed in the use and limitations of such equipment, and receive a medical evaluation prior to use.

Powder-Actuated Tools (1926.302). Only employees who have been trained in the operation of the particular tool in use shall be allowed to operate a powder-actuated tool.

Gas Welding and Cutting (1926.350). The employer shall thoroughly instruct employees in the safe use of fuel gas.

Arc Welding and Cutting (1926.351). Employers shall instruct employees in the safe means of arc welding and cutting.

Fall Protection (1926.503). The employer shall provide a training program for each employee who might be exposed to fall hazards. The program shall enable each employee to recognize the hazards of falling and shall train each employee in the procedures to be followed in order to minimize these hazards.

Site Clearing (1926.604). Employees engaged in site clearing shall be protected from hazards of irritant and toxic plants and suitably instructed in the first aid treatment available.

Stairways and Ladders (1926.1060). The employer shall provide a training program for each employee using ladders and stairways, as necessary. The program shall enable each employee to recognize hazards related to ladders and stairways, and shall train each employee in the procedures to be followed to minimize these hazards.

IMPLIED TRAINING REQUIREMENTS

General Safety and Health Provisions (1926.20). Such programs [as may be necessary to comply with this part] shall provide for frequent and regular inspections of the jobsites, materials, and equipment to be made by competent persons [who are capable of identifying existing and predictable hazards in the surroundings or working conditions that are unsanitary, hazardous, or dangerous to employees, and who have authorization to take prompt corrective measures to eliminate them designated by the employers].

The employer shall permit only those employees qualified by training or experience to operate equipment and machinery.

Gases, Vapors, Fumes, Dusts, and Mists (1926.55). To achieve compliance with paragraph (a) of this section, administrative or engineering controls must first be implemented whenever feasible. When such controls are not feasible to achieve full compliance, protective equipment or other protective measures shall be used to keep the exposure of employees to air contaminants within the limits prescribed in this section. Any equipment and technical measures used for this purpose must first be approved for each particular use by a competent industrial hygienist or other technically qualified person. Whenever respirators are used, their use shall comply with 1926.103.

Hearing Protection (1926.101). Ear protective devices inserted in the ear shall be fitted or determined individually by competent persons.

Ground-Fault Protection (1926.404). The employer shall designate one or more competent persons to implement the program.

Cranes and Derricks (1926.550). The employer shall designate a competent person who shall inspect all machinery and equipment prior to each use, and during use, to make sure it is in safe operating condition.

Excavation and Trenching (1925.651). Daily inspections of excavations, the adjacent areas, and protective systems shall be made by a competent person for evidence of a situation that could result in possible cave-ins, indications of failure of protective systems, hazardous atmospheres, or other hazardous conditions. An inspection shall be conducted by the competent person prior to the start of work and as needed throughout the shift.

Blasting and Use of Explosives (1926.900). The employer shall permit only authorized and qualified persons to handle and use explosives.

Ladders (1926.1053). Ladders shall be inspected by a competent person for visible defects on a periodic basis and after any occurrence that could affect their safe use.

ADOPTED TRAINING REQUIREMENTS

Fire Protection (1926.150). Portable fire extinguishers shall be inspected periodically and maintained in accordance with Maintenance and Use of Portable Fire Extinguishers, NFPA No. 10A-1970. From ANSI Standard 10A-1970: The name plate(s) and instruction manual should be read and thoroughly understood by persons who may be expected to use extinguishers.

Woodworking Tools (1926.304). All woodworking tools and machinery shall meet other applicable requirements of the ANSI Standard 01.1-1961. From ANSI Standard 01.1-1961: Before a worker is permitted to operate any woodworking machine, he [or she] shall receive instructions in the hazards of the machine and the safe method of its operation.

Material Handling Equipment (1926.602). All industrial trucks in use shall meet the applicable requirements of design, construction, stability, inspection, testing, maintenance, and operation, as defined in ANSI B56.1-1969, Safety Standards for Powered Industrial Trucks. From ANSI Standard B56.1-1969: Only trained and authorized operators shall be permitted to operate a powered industrial truck.

NEW HIRE ORIENTATION

Your new hire orientation program should guide the new employee though the basics of your company's safety program. Employers often take for granted that new hires have been previously trained or that they just know how to do the job. Although that may be the case, it never hurts to offer a basic refresher training program.

The training should cover safety requirements, procedures for handling and reporting unsafe situations, emergency action plans, and your hazardous communication program. See Appendix C for a sample new hire orientation checklist.

Companies often use safety videos and handbooks to supplement their new hire orientation programs. NAHB Builder-Books offers several such items, including the *NAHB-OSHA Jobsite Safety Handbook* and the *Jobsite Safety Video*, which can be purchased online at www.BuilderBooks.com.

RECORDING TRAINING HISTORY

Be sure to keep records of your training sessions. Topics like first aid and CPR need to be updated regularly, so good recordkeeping is essential. Employees should be required to sign an attendee sheet to verify their participation. This information could prove helpful in a lawsuit or an OSHA inspection.

It is also a good idea to keep a copy of employee certifications in the employee's personnel file. Employers should also maintain a comprehensive checklist of employee training (Table 2). This will allow you to easily determine which employees require new or refresher training sessions.

ONGOING TRAINING AND AWARENESS

The future of home building involves many new innovations. The evolution of safety has just begun. As home building projects get more complex, you will need to be diligent in monitoring and updating your safety measures to ensure that your employees can build a better tomorrow, in the confines of a safe jobsite.

Ongoing training is a key component to keeping your safety program on track. Weekly toolbox talks are an effective training method. When performed properly, these brief subject-specific training sessions can help your employees maintain a high level of safety awareness. You can develop your own toolbox talks, or purchase the *NAHB Tool Box Safety Talks*, available through BuilderBooks. Other forms of ongoing training include online courses and local training classes.

Table 2. Sample Employee Training Record

Employee Name	First-Aid CPR	OSHA 10 Hour	HazCom	Scaffold
Eric Adams	4-14-05	2-18-06		
Jane Brown				9-14-05
Sally Johnson			6-11-05	
John Smith	2-11-03	3-19-04	6-11-05	7-14-06

APPENDIXES

Appendix A

Additional Sources of Information

- National Institute for Occupational Safety and Health (NIOSH) www.cdc.gov/Niosh/homepage.html

- American National Standards Institute (ANSI): www.ansi.org

- National Fire Protection Association (NFPA): www.nfpa.org

- National Association of Home Builders (NAHB): www.nahb.org

The following publications can be downloaded from OSHA's Web site (www.osha.gov):

- Asbestos Standard for the Construction Industry: www.osha.gov/Publications/OSHA3096/3096.html

- A Guide for Protecting Workers from Woodworking Hazards: www.osha.gov/Publications/osha3157.pdf

- A Guide to Scaffold Use in the Construction Industry: www.osha.gov/Publications/osha3150.pdf

- Chemical Hazard Communication: www.osha.gov/Publications/osha3084.pdf

- Controlling Electrical Hazards: www.osha.gov/Publications/3075.html

- Employer Rights and Responsibilities Following an OSHA Inspection: www.osha.gov/Publications/osha3000.html

- Excavations: www.osha.gov/Publications/OSHA2226/2226.html

- Fall Protection in Construction: www.osha.gov/Publications/3146.html

- Hand and Power Tools: www.osha.gov/Publications/osha3080.html

- Hazard Communication Guidelines for Compliance: www.osha.gov/Publications/osha3111.html

- Personal Protect Equipment: www.osha.gov/Publications/osha3151.pdf

- Training Requirements in OSHA Standards and Training Guidelines: www.osha.gov/Publications/2254.html

- Selected Construction Regulations for the Home Building Industry: www.osha.gov/Publications/Homebuilders/Homebuilders.html

- Stairways and Ladders: A Guide to OSHA Rules: www.osha.gov/Publications/osha3124.pdf

- Worker Safety Series: www.osha.gov/Publications/OSHA3252/3252.html

APPENDIX B

OSHA Regional Offices and States with Approved Programs

REGIONAL OFFICES

These states and territories operate their own OSHA-approved job safety and health programs (Connecticut and New York plans cover public employees only).

Region I
(CT,* MA, ME, NH, RI, VT*)
JFK Federal Building, Room E-340
Boston, MA 02203
(617) 565-9860

Region II
(NJ, NY,* PR,* VI*)
201 Varick Street, Room 670
New York, NY 10014
(212) 337-2378

Region III
(DC, DE, MD,* PA, VA,* WV)
The Curtis Center, Suite 740 West
170 S. Independence Mall
West Philadelphia, PA 19106-3309
(215) 861-4900

Region IV
(AL, FL, GA, KY,* MS, NC,* SC,* TN*)
Atlanta Federal Center
61 Forsyth Street, SW, Room 6T50
Atlanta, GA 30303
(404) 562-2300

Region V
(IL, IN,* MI,* MN,* OH, WI)
230 South Dearborn Street
Room 3244
Chicago, IL 60604
(312) 353-2220

Region VI
(AR, LA, MN,* OK, TX)
525 Griffin Street, Room 602
Dallas, TX 75202
(972) 850-4145

Region VII
(IA,* KS, MO, NE)
City Center Square
1100 Main Street, Suite 800
Kansas City, MO 64105
(816) 426-5861

Region VIII
(CO, MT, ND, SD, UT,* WY*)
1999 Broadway, Suite 1690
Denver, CO 80802-5716
(720) 264-6550

Region IX
(American Samoa, AZ,* CA,* Guam, HI,* NV,* Trust Territories of the Pacific)
71 Stevenson Street, 4th Floor
San Francisco, CA 94105
(415) 975-4310

Region X
(AK,* ID, OR,* WA*)
1111 Third Avenue, Suite 715
Seattle, WA 98101-3212
(206) 553-5930

STATES WITH APPROVED PLANS

Alaska
Commissioner
Alaska Department of Labor
1111 West 8th Street
Room 306
Juneau, AK 99801
(907) 465-2700

Arizona
Director
Industrial Commission of Arizona
800 W. Washington
Phoenix, AZ 85007
(602) 542-5795

California
Director
California Department of Industrial Relations
1515 Clay St.
Suite 1901
Oakland, CA 94612
(415) 703-5050

Connecticut
Commissioner
Connecticut Department of Labor
200 Folly Brook Boulevard
Wethersfield, CT 06109
(860) 566-5123

Hawaii
Director
Hawaii Department of Labor and Industrial Relations
830 Punchbowl Street
Honolulu, HI 96813
(808) 586-8844

Indiana
Commissioner
Indiana Department of Labor
State Office Building
402 West Washington Street
Room W195
Indianapolis, IN 46204
(317) 232-2378

Iowa
Commissioner
Iowa Division of Labor Services
1000 E. Grand Avenue
Des Moines, IA 50319
(515) 281-8067

Kentucky
Secretary
Kentucky Labor Cabinet
1047 U.S. Highway, 127 South, STE 4
Frankfort, KY 40601
(502) 564-3070

Maryland
Commissioner
Maryland Division of Labor and Industry
Department of Labor, Licensing and Regulation
1100 N. Eutaw Street, Room 613
Baltimore, MD 21201-2206
(410) 767-2241

Michigan
Director
Michigan Department of Consumer and Industry Services
4th Floor, Law Building
P.O. Box 30643
Lansing, MI 48909
(517) 322-1814

Minnesota
Commissioner
Minnesota Department of Labor and Industry
443 Lafayette Road North
St. Paul, MN 55155
(651) 284-5010

Nevada
Administrator
Nevada Division of Industrial Relations
400 West King Street
Carson City, NV 89710
(702) 486-9020

New Jersey
Office of Public Employees Occupational Safety and Health
1 John Fitch Plaza
P.O. Box 386
Trenton, NJ 08625-0386
(609) 292-2975

New Mexico
Secretary
New Mexico Environment Department
1190 St. Francis Drive
PO Box 26110
Santa Fe, NM 87502
(505) 827-2850

New York
Commissioner
New York Department of Labor
W. Averell Harriman State Office Building - 12, Room 500
Albany, NY 12240
(518) 457-2741

North Carolina
Commissioner
North Carolina Department of Labor
4 West Edenton St.
Raleigh, NC 27601
(919) 807-2900

Oregon
Administrator
Department of Consumer & Business Services
Occupational Safety and Health Division (OR-OSHA)
350 Winter Street, NE, Room 430
Salem, OR 97301
(503) 378-3272

South Carolina
Director
South Carolina Department of Labor, Licensing and Regulation
Koger Office Park, Kingstree Building
110 Centerview Drive
PO Box 11329
Columbia, SC 29210
(803) 896-4300

Tennessee
Commissioner
Tennessee Department of Labor
710 James Robertson Parkway
Nashville, TN 37243-0659
(615) 741-2793

Utah
Commissioner
Industrial Commission of Utah
160 East 300 South, 3rd Floor
PO Box 146650
Salt Lake City, UT 84114-6650
(801) 530-6898

Vermont
Commissioner
Vermont Department of Labor
    and Industry
National Life Building—
    Drawer 20
National Life Dr.
Montpelier, VT 05620-3401
(802) 828-4301

Virginia
Commissioner
Virginia Department of Labor
    and Industry
Powers-Taylor Building
13 South 13th Street
Richmond, VA 23219
(804) 786-2377

Washington
Director
Washington Department of
    Labor and Industries
General Administrative Building
PO Box 44001
Olympia, WA 98504-4001
(360) 902-4200

Wyoming
Administrator
Worker's Safety and
    Compensation Division (WSC)
Wyoming Department of
    Employment
Cheyenne Bus. Center
1510 E. Pershing Blvd.
Cheyenne, WY 82002
(307) 777-7786

Puerto Rico
Secretary
Puerto Rico Department of Labor
    and Human Resources
Prudencio Rivera Martinez
    Building
505 Munoz Rivera Avenue
Hato Rey, PR 00918
(787) 754-2119

Virgin Islands
Commissioner
Virgin Islands Department of
    Labor
2203 Church St.
Christiansted
St. Croix, VI 00820-4666
(809) 773-1994

AREA OFFICES

Area	Telephone
Albany, NY	(518) 464-4338
Albuquerque, NM	(505) 248-5302
Allentown, PA	(610) 776-0592
Anchorage, AK	(907) 271-5152
Appleton, WI	(920) 734-4521
Augusta, ME	(207) 626-9160
Austin, TX	(512) 374-0271
Avenel, NJ	(908) 750-3270
Bangor, ME	(207) 941-8177
Baton Rouge, LA	(225) 298-5458
Bayside, NY	(718) 279-9060
Bellevue, WA	(425) 450-5480
Billings, MT	(406) 247-7494

Area	Telephone
Birmingham, AL	(205) 731-1534
Bismarck, ND	(701) 250-4521
Boise, ID	(208) 321-2960
Bowmansville, NY	(716) 684-3891
Braintree, MA	(617) 565-6924
Bridgeport, CT	(203) 579-5581
Calumet City, IL	(708) 891-3800
Carson City, NV	(775) 687-5240
Charleston, WV	(304) 347-5937
Cincinnati, OH	(513) 841-4132
Cleveland, OH	(216) 615-4266
Columbia, SC	(803) 765-5904
Columbus, OH	(614) 469-5582

Area	Telephone
Concord, NH	(603) 225-1629
Corpus Christi, TX	(361) 888-3420
Dallas, TX	(214) 320-2400
Denver, CO	(303) 844-5285
Des Plaines, IL	(847) 803-4800
Des Moines, IA	(515) 284-4794
Eau Claire, WI	(715) 832-9019
El Paso, TX	(915) 534-6251
Englewood, CO	(303) 843-4500
Erie, PA	(814) 461-1492
Fairview Heights, IL	(618) 632-8612
Fort Lauderdale, FL	(954) 424-0242
Fort Worth, TX	(817) 428-2470
Frankfort, KY	(502) 227-7024
Guaynabo, PR	(787) 277-1560
Harrisburg, PA	(717) 782-3902
Hartford, CT	(860) 240-3152
Hasbrouck Heights, NJ	(201) 288-1700
Honolulu, HI	(808) 586-8844
Houston, TX	(281) 286-0583
Houston, TX	(281) 591-2438
Indianapolis, IN	(317) 226-7290
Jackson, MS	(601) 965-4606
Jacksonville, FL	(904) 232-2895
Kansas City, MO	(816) 483-9531
Linthicum, MD	(410) 865-2055
Little Rock, AR	(501) 224-1841
Lubbock, TX	(806) 472-7681
Madison, WI	(608) 441-5388
Marlton, NJ	(609) 757-5181
Methuen, MA	(617) 565-8110
Milwaukee, WI	(414) 297-3315
Minneapolis, MN	(612) 664-5460
Mobile, AL	(251) 441-6131
Nashville, TN	(615) 781-5423
New York, NY	(212) 620-3200
Norfolk, VA	(757) 441-3820
North Aurora, IL	(630) 896-8700
Oakland, CA	(415) 703-5050
Oklahoma City, OK	(405) 278-9560
Omaha, NE	(402) 553-0171
Parsippany, NJ	(973) 263-1003
Peoria, IL	(309) 589-7033
Philadelphia, PA	(215) 597-4955
Phoenix, AZ	(602) 542-4411
Pittsburgh, PA	(412) 395-4903
Portland, OR	(503) 326-2251
Providence, RI	(401) 528-4669
Raleigh, NC	(919) 790-8096
Sacramento, CA	(916) 263-5765
Salt Lake City, UT	(801) 530-6848
San Antonio, TX	(210) 472-5040
Savannah, GA	(912) 652-4393
Smyrna, GA	(770) 984-8700
Springfield, MA	(413) 785-0123
St. Louis, MO	(314) 425-4249
Syracuse, NY	(315) 451-0808
Tampa, FL	(813) 626-1177
Tarrytown, NY	(914) 524-7510
Toledo, OH	(419) 259-7542
Tucker, GA	(770) 493-6644
Westbury, NY	(516) 334-3344
Wichita, KS	(316) 269-6644
Wilkes-Barre, PA	(717) 826-6538
Wilmington, DE	(302) 573-6518

Appendix C
Forms and Checklists

88 Employee Safety Violation Reprimand Form

89 Accident Investigation Report Form

91 Construction Safety Checklist

94 New Hire/Trade Contractor Safety Orientation Checklist

95 Safety Program Checklist

EMPLOYEE SAFETY VIOLATION REPRIMAND FORM

Employee name: _____ Date: _____

Location: _____

Safety violation—Circle appropriate number(s):

1. Housekeeping/sanitation
2. Personal protective equipment
3. Ladders/scaffolding
4. Portable power & hand tools
5. Powder-actuated tools
6. Welding & cutting
7. Materials handling & storage
8. Fall protection
9. Excavation/trenching
10. Electrical
11. Fire protection
12. Concrete/masonry
13. Other

Safety violation description/explanation (use additional sheet if necessary)

_____

_____

_____

Supervisor _____

Signature _____ Date _____

Employee response

_____

_____

_____

Signature _____ Date _____

I acknowledge receipt of this safety violation reprimand form. My signature does not imply that I agree with the content.

Employee signature _____

ACCIDENT INVESTIGATION REPORT FORM

Project name _____ Date_____

Exact location_____

Personal Injury

Name of injured person _____

Address _____

Occupation _____

Injury _____

Treatment rendered by _____

Nature of injury _____

Object/equipment/substance inflicting injury _____

Person with most control of object/equipment/substance _____

Property Damage

Property damaged _____

Estimated costs $_____ Actual costs $_____

Owner of property_____

❏ Trade contractors   ❏ Clients   ❏ Company   ❏ Others

Estimated replacement time _____

Nature of damage _____

Object/equipment/substance inflicting injury _____

Person with most control of object/equipment/substance _____

Continued

ACCIDENT INVESTIGATION REPORT FORM *Continued*

Analysis

Describe how the accident occurred. (Attach diagram for all motor vehicle accidents.)

_____
_____
_____
_____

What acts, failures to act and/or conditions contributed most directly to this accident?

_____
_____
_____
_____

What are the basic or fundamental reasons for the existence of these acts and/or conditions?

_____
_____
_____
_____

Loss severity potential: ❏ Major ❏ Serious ❏ Minor
Probable recurrence rate: ❏ Frequent ❏ Occasional ❏ Seldom

Prevention

What action has or will be taken to prevent recurrence? Place an X by items that have been completed.

Action	Completed

Interviewed by _____ Date_____

Reviewed by _____ Date_____

CONSTRUCTION SAFETY CHECKLIST

Housekeeping and Sanitation

- ❏ General condition of work areas acceptable
- ❏ Adequate trash removal
- ❏ Floor openings covered or guarded properly
- ❏ Stairs and walkways cleared of debris and materials
- ❏ Note any slip, trip, or fall hazards; guardrails erected on stairways, wall openings, etc.
- ❏ Adequate lighting
- ❏ Adequate ventilation
- ❏ Toilet facilities adequate
- ❏ Drinking water and cups provided

Personal Protective Equipment

- ❏ Hard hats
- ❏ Eye and face protection
- ❏ Gloves
- ❏ Respirators
- ❏ Hearing protection

Ladders and Scaffolding

- ❏ In good, serviceable condition
- ❏ Properly positioned and secured at the top
- ❏ Extend 36 in. above roof or platform, if used for access
- ❏ Doors blocked open, locked or guarded off if in front of ladder
- ❏ Stepladders fully open when used
- ❏ Metal or conductive ladders not used for work within 10 ft. of energized electrical lines or equipment
- ❏ Sound, rigid footing for all scaffolds
- ❏ Safe access to all working levels
- ❏ Equipped with standard guardrails; top rails, midrails, and toe-boards
- ❏ Protection provided where persons are required to work or pass under scaffolding in use
- ❏ No accumulation of tools or material on platforms
- ❏ Outriggers installed, if required
- ❏ Self-propelled (motorized) types of scaffolds require special maintenance and inspection

Portable Power and Hand Tools

- ❏ General condition of tools acceptable
- ❏ Proper tool being used for job being performed
- ❏ Guards and safety devices are operable and in place
- ❏ Electrical tools inspected and marked according to the Assured Equipment Grounding Program
- ❏ Tool retainers used on pneumatic tools; air pressure properly regulated
- ❏ Check for pinch and shear points

Powder-Actuated Tools

- ❏ All operators trained and certified

Continued

CONSTRUCTION SAFETY CHECKLIST *Continued*

Powder-Actuated Tools *Continued*

- ❒ Tools and charges protected from unauthorized use
- ❒ Loaded tools are not left unattended
- ❒ All tools inspected and tested daily before use
- ❒ Tools and charges matched to recommended materials only
- ❒ Safety glasses or face shields used by operators
- ❒ Tools are in compliance with local regulations

Welding and Cutting

- ❒ Operators trained and qualified
- ❒ Personal protective equipment
- ❒ Fire extinguishers provided
- ❒ Flammable/combustible materials protected
- ❒ Gas cylinders secured
- ❒ All fittings free of oil and grease
- ❒ Flashback protection used
- ❒ Regulators and gauges operating
- ❒ Proper gauge settings
- ❒ All hoses, cords, and other equipment in good condition

All Material Storage and Handling

- ❒ Materials properly stacked and on firm footings; properly blocked and secured
- ❒ Fire protection adequate
- ❒ All rigging and lifting equipment properly maintained and periodically inspected
- ❒ Employees picking up and handling loads properly
- ❒ Flammable liquids stored only in approved containers
- ❒ Flammable gasses properly stored
- ❒ Adequate security measures are in place

Fall Protection

- ❒ Guardrails provided where necessary
- ❒ Personal Fall Arrest Systems (PFAS) provided where necessary
- ❒ Fall protection plan implemented where PFAS not used

Excavation and Trenching

- ❒ All excavations/trenches properly shored
- ❒ All excavations have proper means of entry/exit
- ❒ Excavations/trenches inspected by competent person

First Aid Kits

- ❒ Kits provided
- ❒ Kits inspected and replenished where necessary

Hazard Com

- ❒ HazCom rule reviewed

CONSTRUCTION SAFETY CHECKLIST *Continued*

Hazard Com *Continued*

- ❐ Written hazard communication program prepared
- ❐ Employee has been designated to coordinate the written hazard communication program
- ❐ System for responding to medical requests for chemical information during a medical emergency in place
- ❐ Emergency procedures developed and incorporated into standard operating procedures
- ❐ Policy for exchanging and documenting hazard information with other contractors developed
- ❐ Existing company policies and operating procedures reviewed HazCom compliance and amended where necessary
- ❐ Inventory list of all chemicals used in the workplace is on file
- ❐ List of all hazardous chemicals on the jobsite is available
- ❐ Procedure for checking MSDS receipt from the manufacturer for incoming and updated chemicals established
- ❐ Incoming container labels are reviewed to verify the chemical's name and appropriate hazard warnings
- ❐ Labels, placards, or batch tickets for use when products are removed from their original containers or mixed developed
- ❐ System for communicating the information in MSDSs to all company employees has been developed
- ❐ Procedures for employee training on chemical hazards at initial assignment and whenever new information becomes available have been developed
- ❐ Employees are familiar with the different types of chemicals and the hazards associated with them
- ❐ Employees understand how to detect the presence or release of hazardous chemicals in the workplace
- ❐ Employees are trained in the proper work practices and personal protective equipment in relation to the hazardous chemicals in their work area
- ❐ Training programs provide information on appropriate first aid, emergency procedures, and likely symptoms of overexposure
- ❐ Training programs include explanations of labels and warnings that are used in each work area
- ❐ Training details where to obtain MSDSs and how to use them
- ❐ A recordkeeping system to show that employees have been informed of hazards exists
- ❐ Employees who are not routinely exposed to chemicals know where to find the MSDS forms and how to use them

Project name _____ Project # _____

Superintendent_____ Date_____

NEW HIRE/TRADE CONTRACTOR SAFETY ORIENTATION CHECKLIST

Job # _____ Description _____

Project name _____ Phone _____

Address _____

Job superintendent _____ Work start _____

Date _____ Time _____ Place _____

Attendees _____
_____

Subjects covered at meeting:

- ❒ Safety, health policies, and procedures
- ❒ Confined space entry procedure
- ❒ Personal protective equipment required
- ❒ Nonpotable construction water
- ❒ Medical/first aid safety rules
- ❒ Site cleanup/trash disposal
- ❒ Employee conduct on job
- ❒ Fire prevention/protection
- ❒ Insurance coverage
- ❒ Federal/state OSHA standards

- ❒ Vehicle passes
- ❒ Lockout/tag out procedures
- ❒ Temporary power for trade contractor use
- ❒ Drinking water
- ❒ Emergency procedures
- ❒ Accident/injury reporting
- ❒ Workers' comp first reports of injury
- ❒ Environmental/pollution control
- ❒ Other _____
  _____

Remarks _____
_____
_____
_____

SAFETY PROGRAM CHECKLIST

Use this checklist to ensure that you have included all of the necessary components in your safety program.

- ❐ Stated goals in writing
- ❐ Wrote an action plan
- ❐ Established a budget
- ❐ Designated a "safety champion"
- ❐ Defined areas of responsibility
- ❐ Developed and implemented jobsite safe work practices
- ❐ Established accountability procedures
- ❐ Developed a safety training program
- ❐ Developed enforcement procedures
- ❐ Conducted regular jobsite inspections and hazard analysis
- ❐ Developed recordkeeping procedures
- ❐ Established accident reporting and investigation requirements
- ❐ Developed a trade contractor safety compliance policy
- ❐ Evaluated the effectiveness of your safety program
- ❐ Encouraged feedback and rewarded excellence

Appendix D

Model Safety Program

This model safety program should be used only as a guide. Because individual company circumstances vary widely, you should modify the program to reflect actual company operations. This program does not contain legal advice. You should consult with your attorneys or insurance company representatives before acting on the premises described herein.

OUR SAFETY PROGRAM

Mission

Safety is as critical to our operations as planning, scheduling, or billing. We believe accidents are preventable. Therefore, we are all responsible for ensuring that jobsite safety practices are a routine part of our daily work.

Our primary goal is to protect our employees. This safety program is intended to control and prevent those construction jobsite failures that cause fatalities, injuries, illness, equipment damage, or fire and that cause damage to or destruction of property at the jobsite.

Our policy is to strive for the safest possible performance on each of our jobsites. The following safety and loss-control guidelines represent a wealth of practical experience that has been tested in the safety-conscious environment of many successful projects. These procedures are designed to protect our employees and resources from any harm or financial loss caused by accidents. Therefore, as a condition of your employment, you are required to understand and abide by these procedures.

This written safety and total loss-control program may be revised periodically. All employees are encouraged to make suggested revisions to this safety program. Please forward your written comments and suggestions to the president/owner.

RESPONSIBILITIES

You may need to modify the following responsibilities to reflect specific company operations and the actual number of personnel positions in the company.

President/owner

The president/owner will:

- provide direction and motivation and assign accountability to ensure an active safety and loss-control program for all company construction projects
- establish office and field administration safety and loss-control activities that reflect the company's safety goals and objectives

- establish a budget to fund the safety and loss-control programs
- develop annual safety goals and objectives for site superintendent(s) to meet as part of the superintendent's performance evaluations
- assist the site superintendent in developing site-specific safety and loss-control programs
- ensure that the management team has a working knowledge of all client, governmental, and company safety and loss-control requirements
- participate periodically in various employee safety toolbox presentations
- review monthly field safety status reports to evaluate each project's safety and insurance performance
- enforce incentive and disciplinary actions necessary to encourage an effective safety program

Depending on the size of your company, some of the following duties may be split between the superintendent and other management personnel, such as a production manager or safety manager. Remodelers may find that assigning the following duties to a lead carpenter or other lead employee is more practical.

Site superintendent

The site superintendent is responsible for the safety of all company field employees on the project. The site superintendent will:

- establish comprehensive project safety procedures that comply with applicable client contractual documents and specifications (federal or state OSHA) and company safety and loss-control procedures with the assistance of the president/owner
- monitor the project's safety status and employee morale by conducting a daily safety inspection of the jobsite and initiating necessary corrective action
- conduct accident investigations, analyze the causes, and recommend corrective and/or preventive actions
- prepare accident reports and maintain documentation of workers' compensation reports
- maintain and update any necessary OSHA records and material safety data sheets (MSDS)
- conduct project safety and loss-control training for employees
- ensure that each jobsite has the necessary safety equipment and materials, personal protective equipment, first aid supplies, and emergency telephone numbers posted
- monitor trade contractor performance to ensure compliance with the company's safety performance requirements
- prepare and distribute job safety bulletins and subject material for toolbox safety meetings and reviews and audit the meetings to ensure effectiveness
- enforce disciplinary actions

- notify the company's president/owner of any serious accident or OSHA inspection as soon as possible
- refine or expand safety procedures as needed to meet the site specific safety and loss-control needs of a particular project

Field employees

As a condition of employment, each employee is expected to work in a manner that will not inflict self-injury or cause injury to others. It is important that all employees understand that they are responsible for their own safety on the job. Therefore, all employees will

- comply with all safety rules and regulations
- immediately report all accidents and injuries to the supervisor
- use the proper tools and personal protective equipment for the job
- report all unsafe conditions to the supervisor
- know the procedures in case of emergencies, including contacting emergency services
- help maintain a safe, clean work area
- participate in the company's safety training program
- set a good example for others to follow

Trade contractors

We expect that our trade contractors will have established their own safety and health programs. Each trade contractor is responsible for the safety of his or her employees on each company project. Trade contractors are expected to

- comply with applicable federal and state OSHA regulations
- supply a copy of his or her company's safety program and MSDS for all materials used on company projects
- immediately report all accidents, injuries, and fatalities that occur on the company jobsite to the company superintendent
- supply the proper personal protective equipment and safety equipment to his or her employees and ensure their use
- adequately train field employees on proper safety practices
- report all unsafe conditions to the site superintendent
- immediately notify the company president/owner or site superintendent in the event of an OSHA inspection when no company personnel are on site

You may need to modify the following accountability procedures to meet your company's specific needs.

ACCOUNTABILITY PROCEDURES

No phase of our operations is of greater importance than accident prevention. Therefore, every employee shall be held accountable for his or her safety and loss-control performance. This accountability will be reflected in retention, promotions, salary increases, bonuses, and perks.

Safety enforcement procedures

The site superintendent will issue a written reprimand on the employee safety violation reprimand form as soon as an infraction has been observed. The reprimand serves to:

- allow employees to change unsafe work practices
- document the infraction
- guarantee that employees are warned of rule infractions prior to further disciplinary action being taken

A reprimand will be issued for the following reasons:

- failure to wear proper protective equipment, such as eye protection
- willfully endangering one's life or the lives of other employees, which is gross misconduct and will be cause for immediate dismissal
- performing work in an unsafe manner

A copy of the form will be maintained in the employee's personnel file.

Disciplinary action

The severity of the discipline will be determined by the extent of the exposure to the employee in question, other employees, and the company. If the incident is the likely cause of an accident or if the violation had a high probability of resulting in an accident, the employee may be terminated. If the incident had a moderate probability of causing an accident, time off without pay may be sufficient. If the incident had a low probability of causing an accident, the employee will receive a written reprimand. Three written reprimands for safety violations will result in immediate termination.

You may need to modify the following reports and recordkeeping procedures to reflect specific company operations and the actual number of personnel positions in the company.

SAFETY REPORTS AND RECORDKEEPING PROCEDURES

The following procedures apply to all company jobsites and will be used to measure the overall safety and insurance performance of each company project.

Administration

The site superintendent can delegate the daily administration of the following reporting and recordkeeping requirements to a staff member. In that event, however, the site superintendent will determine the actual timely and adequate completion and distribution of these reports and records.

General requirements

Requests for forms or records from third parties or external agencies must be approved by the company president/owner. This includes requests from clients and owners of projects.

Forms that have been developed for use at field locations must be approved by the company president/owner prior to being used. Any and all records generated at field locations must be maintained at the location until the project is complete. Safety or medical files or records must not be destroyed.

1. **Company records.** In addition to workers' compensation reports, the site superintendent shall maintain a file of all company safety records, including accident investigation forms, construction safety audits, and trade contractor safety orientation checklists.

Include the following statement if you have employed more than 10 employees, not including trade contractors, during the previous calendar year.

2. **OSHA Log of Work-Related Illnesses and Injuries (Form 300).** Each site superintendent shall be responsible for completing, signing, and maintaining the company's log of accidents and injuries. The log must be retained for five years. In addition the following conditions must be observed:

- When company work from the previous year is still ongoing, the *Summary of Work-Related Illnesses and Injuries* (Form 300A) must be posted on the company jobsite bulletin from February 1 to April 30, after which it may be taken down and filed with other jobsite safety records.
- Under no circumstances will the company site superintendent maintain a log for trade contractors.

You may need to modify the following accident reporting and investigation procedures to reflect specific company operations and the actual number of personnel positions in the company.

ACCIDENT REPORTING AND INVESTIGATION PROCEDURES

Every person in this company, including employees, managers, and owners, is responsible for investigating accidents. We consider an accident to be serious if it results in

- occupational death(s), regardless of the time between injury or illness and death
- occupational illness or illnesses resulting in permanent total disabilities
- occupational accident(s) that involve any property damage
- hospitalizations

RESPONSIBILITIES

You may need to modify the following responsibilities to meet your company's specific needs.

Site superintendent

- ensure that each employee receives prompt first aid treatment for all injuries

- review and correct the causes of all minor injuries to his or her employees
- take any emergency action necessary to minimize the extent of loss to both employees and property when a serious accident occurs
- investigate the accident and report findings and recommendations by completing the accident investigation report form
- immediately notify the company president/owner regarding a serious accident
- complete the appropriate project insurance report forms and forward them to the insurance carrier

President/owner

- provide leadership, guidance, and control to ensure that the accident investigation responsibilities at all levels of site management are effectively administered
- determine whether there is an immediate need to inform the company's legal counsel and insurance agent or broker, based on preliminary information received from the field
- participate in a meeting with the site superintendent and field supervisors to review safety and loss-control policies or procedures that need to be developed or upgraded
- clear all press statements with the company's legal counsel, the client, and the insurance carrier

Note: All statements, with respect to any accident, made to individuals not connected with the company will be handled by the president/owner. Statements that must be made by company field personnel to insurance company representatives or law enforcement authorities will be confined to the basic facts. Further details must be cleared by the company president/owner prior to their release. No statement regarding accident liability will be made to anyone not connected with the company.

GOVERNMENTAL SAFETY AND HEALTH COMPLIANCE

The OSHA Construction Industry Safety and Health Standards (29 CFR 1926) are considered the minimum safety requirements for this company. Therefore, for all company projects the site superintendent will

You may need to modify the following responsibilities to reflect specific company operations.

- obtain copies of the most recent issue of the applicable federal and state OSHA construction safety and health standards
- ensure that OSHA standards are rigorously applied in terms of equipment procedures and job content
- ensure that employees follow OSHA standards by using required equipment and precautions and applying sound principles of employee discipline when employees fail to comply
- maintain appropriate OSHA records

- ensure posting of required notices related to OSHA
- be prepared for and meet the requirements of OSHA inspections

Include the following safe work practices for materials and equipment specific to the jobsite. You may need to modify these safe work practices to meet your company's specific needs.

SAFE WORK PRACTICES FOR PERSONAL PROTECTION EQUIPMENT

All employees must observe the following safe work practices.

- Wear a hard hat when there is a danger from impact, falling or flying objects, or electrical shock.
- Wear impact-resistant safety glasses when you use materials or operate equipment that could result in materials striking your eyes.
- Wear safety goggles if you are working with materials or chemicals that could damage your eyes on contact.
- Wear face shields to protect your face from flying objects or splash hazards.
- Wear proper eye protection when welding, cutting, or brazing.
- Use hearing protection when exposed to hazardous levels of sound.
- Wear proper shoes or boots while on the jobsite to protect against nail puncture injuries.
- Wear respiratory protection when you are exposed to inhalation hazards.

Include the following safe work practices for materials and equipment specific to the jobsite. You may need to modify these safe work practices to meet your company's specific needs.

SAFE WORK PRACTICES FOR FIRE PROTECTION

All employees must observe the following safe work practices.

- Know where the fire extinguishers are located and how to use them.
- Use only approved safety cans for storing more than one gallon of flammable liquid, although the original container may be used for less than one gallon.
- Do not store flammable or combustible liquids in areas used for stairways or exits.
- Do not store combustible materials more than 20 ft. high.
- Keep driveways between and around combustible material piles at least 15 ft. wide.
- Keep areas clean of debris, weeds, and grass.
- Do not store combustible materials within 10 ft. of a building or structure.
- Keep fire extinguishers with a rating of at least 2A within 100 ft. of storage areas.

- Keep fire extinguishers in plain sight and within reach.
- Store flammable liquids in closed containers when not in use.
- Clean up leaks or spills of flammable or combustible liquids promptly.
- Use flammable liquids only where there are no open flames or other ignition sources within 50 ft. of the operation.
- Do not store liquefied petroleum (LP) gas tanks inside buildings.
- Keep LP gas containers with a water capacity greater than 2½ lb. on a firm and level surface and, when necessary, in a secured, upright position.
- Keep temporary heaters at least 6 ft. away from any LP gas container.
- Do not use solid fuel salamanders in buildings or on scaffolds.

Include the following safe work practices for materials and equipment specific to the jobsite. You may need to modify these safe work practices to meet your company's specific needs.

SAFE WORK PRACTICES FOR TOOLS

All employees must observe the following safe work practices.

General guidelines

- Maintain all hand and power tools (employee or employer owned) in safe condition.
- Follow the manufacturer's requirements for safe use of all tools.
- Inspect all hand and power tools before use. If a tool is found to be defective, remove it from service and notify the supervisor.
- Keep guards on tools at all times.
- Use all required personal protective equipment when using tools.
- Do not use wrenches when the jaws are sprung to the point of slippage.
- Do not use impact tools with mushroomed heads.
- Keep wooden handles free of splinters and cracks and ensure that they fit securely in the tool.
- Never point tools at anyone.
- Test the safety device on the tool each day before using.
- Do not leave loaded tools unattended.
- Use personal protective equipment that meets OSHA requirements.
- Ask your supervisor for help or training if you need it.

Power tools

- Keep all electric power tools grounded or use the double-insulated type.
- Do not use electric cords or hoses for hoisting or lowering tools.
- Secure pneumatic power tools to the hose or whip to prevent them from accidentally disconnecting.

- Use safety clips or retainers on pneumatic impact tools to prevent attachments from accidentally expelling.
- Do not use compressed air for cleaning, except when below 30 psi and only with personal protective equipment. The 30-psi requirement does not apply to concrete form, mill scale, and similar cleaning purposes.
- Work within the manufacturer's safe operating pressure for hoses, pipes, valves, filters, and other fittings.

Woodworking tools

- Use only portable, power-driven circular saws that have guards above and below the base plate or shoe. When the tool is withdrawn from work, the lower guard must automatically and instantly return to the covering position.
- Maintain guards on all the moving parts and blades of other portable saws and equipment.
- Load tools only just before use.
- Do not fire into materials that are too hard or too soft.
- Reinforce materials if there is a chance that fasteners could go through the material.
- Do not try to fasten an area that is spoiled from other attempts to fasten.

Powder-actuated tools

- Only operate a powder-actuated tool if you have been trained in the operation of the particular tool.
- Test the tool before loading each day in accordance with the manufacturer's recommended procedure to ensure that the safety devices are in proper working condition.
- Do not load tools until just before the intended firing time.
- Keep hands clear of the open barrel end.
- Do not drive fasteners into very hard or brittle material.

Include the following safe work practices for materials and equipment specific to the jobsite. You may need to modify these safe work practices to meet your company's specific needs.

SAFE WORK PRACTICES FOR WELDING AND CUTTING

All employees must observe the following safe work practices.

Transporting, moving, and storing compressed gas cylinders

- Protect valves of stored cylinders.
- Secure cylinders on a cradle, sling board, or pallet when hoisting.
- Do not hoist cylinders with magnets or choker slings.
- Move cylinders short distances by tilting and rolling them on the bottom edges, not the sides.
- Do not drop cylinders or allow them to collide.

- Secure cylinders in an upright position during transport.
- Do not lift cylinders by the valve caps.
- Do not use bars to pry frozen valves or valve caps loose.
- Secure valve caps before moving cylinders unless they are secure in a cylinder carrier.
- Secure cylinders in a truck or cart to prevent them from falling or being knocked over while in use.
- Do not use cylinders as rollers or supports.
- Close all valves before moving cylinders.
- Secure cylinders in an upright position except when briefly hoisting or carrying them.
- Separate oxygen cylinders in storage from fuel/gas cylinders or combustible materials cylinders (especially oil or grease). This separation must be either a minimum distance of 20 ft. or a non-combustible barrier at least 5 ft. high with a ½-hour fire-resistance rating.
- Store cylinders in a well-protected, well-ventilated, dry location at least 20 ft. from highly combustible materials such as oil or excelsior.
- Do not store cylinders in unventilated enclosures such as lockers and cupboards.
- Protect stored cylinders from tampering and damage.

Usage

- Crack the valve before connecting a regulator by opening the valve slightly and then immediately closing it.
- Stand on the side of outlets when cracking the valve.
- Do not crack fuel gas cylinders near welding work, sparks, flame, or other ignition sources.
- Open the valve slowly to prevent damage to the regulator.
- Do not open quick-closing valves by more than 1½ turns.
- Leave any needed wrenches in position on the stem of the valve while the cylinder is in use so that the fuel gas flow can be shut off quickly if necessary.
- Do not place anything on top of a cylinder in use that may damage the safety device or interfere with the quick closing of the valve.
- Do not use fuel gas from cylinders through torches or other devices without a pressure-reduction regulator attached to the cylinder valve.
- Close the valve and release gas from the regulator before removing the regulator from the cylinder valve.
- Close the valve and tighten the gland nut of full gas cylinders when a leak is found around the valve stem.
  - If this action does not stop the leak, disconnect the cylinder and remove it

from service, tag it, and remove it from the work area.

- If fuel gas leaks from the cylinder valve rather than from the valve stem and the gas cannot be shut off, tag the cylinder and remove it from the work area.

- If a regulator attached to a cylinder valve stops a leak through the valve seat, the cylinder does not need to be removed from the work area.

- If a leak develops at a fuse plug or other safety device, remove the cylinder from the work area.

- Use fire shields or keep cylinders far enough away from welding or cutting operations so that sparks, hot slag, or flame cannot reach them.

- Keep cylinders away from energized electrical circuits.

- Do not strike a cylinder with an electrode to strike an arc.

- Use fuel gas cylinders in an upright position only.

- Keep fuel gas cylinders away from flame, hot metal, or other sources of heat.

- Keep cylinders containing oxygen, acetylene, or other fuel gas out of confined spaces.

- Keep a fire extinguisher in the work area of welding, cutting, or flames.

- Remove damaged or defective cylinders from the work place.

- Do not use oxygen for ventilation, comfort cooling, blowing dust from clothing, or for cleaning work areas.

- Keep oxygen cylinders and fittings away from oil and grease.

- Keep cylinders, caps and valves, couplings, regulator hoses, and apparatus free of oil or greasy materials and don't handle them with oily hands or gloves.

- Maintain ventilation during operations.

- Use fans, open windows, or work outdoors to keep fume levels low.

Hoses

- Make sure you can differentiate between fuel gas and oxygen hoses. Use different colors or surface characteristics. Oxygen and fuel gas hoses are not interchangeable.

- Do not use a single hose that has more than one gas passage.

- Do not cover more than 4 of 12 in. when taping parallel sections of oxygen and fuel gas hose together.

- Inspect all hoses that will carry any gas or substance that may ignite or be harmful.

- Remove defective and damaged hoses from service.

- Use double-locking or rotary connections for hoses.

- Ventilate storage areas and boxes used to store gas hoses.

- Keep passageways, ladders, and stairs clear of hoses and other equipment.

Torches

- Clean clogged torch tip openings with suitable cleaning wires or drills.
- Inspect torches before use for leaking shutoff valves, hose couplings, and tip connections.
- Do not use defective torches.
- Use friction lighters to light torches.

Include the following safe work practices for materials and equipment specific to the jobsite. You may need to modify these safe work practices to meet your company's specific needs.

SAFE WORK PRACTICES FOR ELECTRICITY

All employees must observe the following safe work practices.

General usage

- Do not contact any electrical power circuit unless the circuit is de-energized or guarded by insulation or other means.
- Wear insulated gloves when using jackhammers, bars, or other hand tools that may contact a line when the underground location of the power lines is unknown.
- Remove all damaged electrical tools from service.
- Protect electrical equipment from contact in passageways.
- Keep all walking/working surfaces free of electrical cords.
- Do not use worn or frayed electrical cords or cables.
- Do not fasten extension cords with staples.
- Do not hang cords from nails or suspend them with wire.
- Maintain a minimum of 10 ft. from all energized power lines.

Ground fault protection and temporary power

- Use ground-fault circuit interrupters (GFCI) to protect extension cords and any other connectors even if the cords are connected to the permanent wiring of the house.
- Install all temporary receptacles in complete metallic raceways.
- Protect all general lighting lamps from breakage.
- Ground all metal case sockets.
- Protect extension cords that are run through doors, windows, and floor holes.
- Use only three-wire type extension cords designed for hard or junior hard service. Look for the following letters imprinted on the casing: S, ST, SO, STO, SJ, SJT, SJO, or SJTO.
- Do not bypass any protective system or device designed to protect you from contact with electrical current.

MODEL SAFETY PROGRAM

Lockout/tag out

- Controls that are to be deactivated during the course of work on energized or de-energized equipment or circuits must be tagged or marked.

- Equipment or circuits that are de-energized must be rendered inoperative.

- Attach tags at all points where equipment or circuits can be energized.

- The tags should be placed to clearly identify which pieces of equipment or circuits are being serviced.

Include the following safe work practices for materials and equipment specific to the jobsite. You may need to modify these safe work practices to meet your company's specific needs.

SAFE WORK PRACTICES FOR SCAFFOLDS

All employees must observe the following safe work practices.

General guidelines

- Wear a hard hat anytime you are working on or near scaffolds.

- Build all scaffolds according to the competent person and manufacturer's directions.

- Ensure that each scaffold is strong enough to support the platform and at least four times the expected load.

- Build all working-level scaffold platforms at least 18 in. wide.

- Deck the platform with no more than a 1 in. space between the decking/platform units and the upright supports. If there is not enough space to fully plank/deck, then you must plank/deck as much as possible.

- Deck as much as necessary to protect yourself when using the platform as a walkway, or for employees who will be erecting or dismantling the scaffold.

  - *Exception:* The decking/platforms for ladder-jack, pump-jack, top-plate, and roof bracket scaffolds can be as narrow as 12 in. wide.

- Make the decking as wide as possible if there is not enough space to build the minimum platform size.

- Keep the front edge of the platform within 14 in. of the face of the work. If this is not possible, you must use guardrails or PFASs to keep from falling to the inside of the work area.

  - *Exception:* The distance between the edge of the platform and the face of the work can be 18 in. for plastering or lathing.

- Cleat or attach platforms to the scaffold or make the planking extend at least 6 in. past the supports. If a platform is shorter than 10 ft., the platforms must not extend past the supports by more than 12 in. unless there is support for the cantilevered section. Platforms longer than 10 ft. must not extend past the supports by more than 18 in. unless there is support for the cantilevered sections. If you can't access those cantilevered sections, you don't have to support them.

- Build longer platforms with the abutting ends of the plank/deck resting on separate supports, or secure them by some other means.

- Overlap the ends of planks/decking by 12 in. on the supports, or nail or secure the ends together by some other means.

- Do not paint the top or bottom of work platforms with anything that will hide defects. Paint only the sides for identification.

- The competent person must decide if it is safe to intermix scaffold parts.

Access

- Use portable, hook-on, or attachable ladders to access the scaffold when the platform is more than 2 ft. above or below the access point. You can also have direct access from another scaffold or the actual structure as long as it is not more than 14 in. away.

- Don't use cross braces to climb on or off scaffolds.

- Place portable, hook-on, and attachable ladders securely to prevent the scaffold from tipping. Make sure that the bottom rung is no more than 24 in. above the ground or floor.

- Use the proper ladder to access the scaffold you are using. Rungs must be at least 12 in. long with maximum spacing of 12 in. between rungs.

Fall protection

- Use a PFAS when working on ladder jacks that are more than 10 ft. above the ground.

- Use a guardrail, PFAS, or grab rope alongside a crawlingboard/chicken ladder.

- Use fall protection (guardrails) on all scaffolds that are more than 10 ft. high.

- Use guardrails along all open sides and ends and build to the following requirements:

  - Top rails between 39 and 45 in. high must be installed.

  - Midrails must be installed halfway between the platform and the top rail. If using mesh or panels, install them from the top to bottom of the guardrail.

  - Guardrails must withstand 200 lb. of downward force and must not be made of steel or plastic banding.

  - Rail ends must not hang over the edge of scaffolds.

  - Midrails and mesh must withstand at least 75 lb. of downward force.

  - Manila or plastic rope can be used as a guardrail only if it is inspected by the competent person and meets the criteria for guardrails.

- Cross bracing can be used in place of top rails or midrails (but not both at the same time) if the cross is between 20 and 30 in. above the platform for the midrail or 38 to 48 in. above the platform for the top rail.
- Surface the guardrails to prevent puncture wounds or lacerations.

Falling object protection

- Wear a hard hat anytime you are working on or near scaffolds.
- Keep objects away from the edge of the scaffolds.
- Build toe boards to a force of at least 50 lb. with a minimum 3 in. in height. If material is taller than the toe boards, netting or other control measures will need to be put into place.

Additional rules for scaffold use

- Do not use any part of a scaffold that is damaged or weakened.
- Do not work on scaffolds if you feel weak, sick, or dizzy.
- Do not work on any part of the scaffold other than the work platform.
- Do not alter the scaffold.
- Do not move a scaffold horizontally while employees are on it unless it is a mobile scaffold and the proper procedures are followed.

- Do not perform heat-producing activities such as welding without taking precautions to protect scaffold components.
- Do not work on scaffolds that are covered with snow, ice, or other slippery matter.
- Do not work on or from scaffolds during storms or high winds unless the competent person has determined that it is safe to do so.
- Do not allow debris to accumulate on platforms.
- Do not overload scaffold platforms.
- Do not use makeshift devices, such as boxes and barrels, on top of scaffold platforms to increase the working level height.
- Do not erect, use, alter, or move scaffolds within 10 ft. of overhead power lines.
- Do not use shore or lean-to scaffolds.
- Do not swing loads near or on scaffolds unless you use a tag line.

Include the following safe work practices for materials and equipment specific to the jobsite. You may need to modify these safe work practices to meet your company's specific needs.

SAFE WORK PRACTICES FOR FALL PROTECTION

All employees must observe the following safe work practices.

Interior falls and guardrails

- Install guardrails or covers whenever there is a fall potential of 6 ft. or more.

- Make guardrail top rails 42 in. ± 3 in. above the walking/working level.

- Raise the top edge height of the top rail equal to the stilt height for an employee who is using stilts.

- Keep midrails halfway between the top edge of the guardrail system and the walking/working level (21 in. high).

- Build guardrail systems to withstand a force of at least 200 lb. (downward or outward thrust) along the top edge.

- Surface the guardrail to prevent injury or clothes snagging.

- Do not use steel banding and plastic banding as top or midrails.

- Erect toe boards along the edge of the walking/working surface.

- Build toe boards to a force of at least 50 lb. with a minimum 3 in. in height. (If material is taller than the toe boards, netting or other control measures will need to be put into place.)

- Build guardrails to the following specs:

  - **For wood railings.** Wood components must be minimum 1,500 lb.–ft./in. (2) fiber (stress grade) construction grade lumber; the posts must be at least 2×4 lumber spaced not more than 8 ft. apart on center; the top rail must be at least 2×4 lumber, the intermediate rail must be at least 1×6 lumber. All lumber dimensions are nominal sizes as provided by the American Softwood Lumber Standards (January 1970).

  - **For pipe railings.** Posts, top rails, and intermediate railings must be at least 1½ in. nominal diameter (schedule 40 pipe) with posts spaced not more than 8 ft. apart on center.

- Guard all unprotected holes with rails or covers.

- Color code or mark the word "hole" or "cover" on the cover.

Controlled access zones (CAZ)

- All access to the CAZ must be restricted to authorized entrants.

- All employees who are permitted in the CAZ must be listed or be visibly identifiable by the competent person before they enter the area.

- All protective elements of the CAZ must be enforced prior to beginning work.

Attic and roof work

- Materials and equipment must be kept in close proximity to the work area.

- Materials and other objects that could be impalement hazards must be kept out of the area below the work space or properly guarded.

- When attic or roof work is in progress, do not stand or walk below or adjacent to any openings in the ceiling where you could be struck by falling objects.

- Operations must be suspended during inclement weather, such as high winds, rain, snow, or sleet that creates a hazardous condition.

Erection of exterior walls

- Attend training before erecting exterior walls.
- Paint a line 6 ft. from the floor deck edge before any wall erection activities to warn of the approaching unprotected edge.
- Stage materials to minimize fall hazards.
- Perform cutting of materials and other preparation as far away from the edge of the deck as possible.

Foundation walls/formwork

- Attend training before you work on the top of the foundation wall/formwork and only as necessary to complete the construction of the wall.
- Ensure that all formwork is adequately supported before you get on top of the formwork.
- Operations must be suspended during inclement weather such as high winds, rain, snow, or sleet.
- Materials and equipment for the work must be kept in close proximity to employees who are on top of the foundation/formwork.
- Materials and other objects that could be an impalement hazard must be properly guarded or kept out of the area below.

Floor joists/trusses and sheathing

- Attend training before you install floor joists or sheathing.
- Stage materials to allow for easy access.
- Roll first floor joists/trusses into position and secure them from ground, ladders, or sawhorse scaffolds.
- Roll each successive floor joist/truss into place and secure them from a platform created from a sheet of plywood laid over the previously secured floor joists or trusses.
  - *Exception:* The first row of sheathing work from the established deck, which must be installed from ladders or the ground.
- If not assisting in the leading-edge construction while leading edges still exist (e.g., cutting the decking for the installers), do not go within 6 ft. of the leading edge under construction.

Roof materials application

Consider the following safe work practices during roofing operations.

- Employees installing shingles and other roofing material should utilize a PFAS during this operation.

- Read all manufacturer's instructions and warnings before using fall protection equipment.

- Anchor points used for personal fall arrest must be capable of supporting 5,000 lb.(2,273 kg) per employee and installed at a secure place on the roof and according to manufacturer's requirements.

- Ensure that the anchor point is located at a height that will not allow the employee to strike a lower level should a fall occur.

- Provide safe ladder access to the roof. If a ladder is used for access, the ladder must extend 3 ft. above the roof line and be secured from movement.

- Load and store roofing materials on the deck in a manner that prevents the materials from sliding off the roof.

- A competent person must determine the proper method of storing material to avoid overloading the roof deck/trusses.

- Provide safe access to the roof level for stocking materials.

- Do not ride a material conveyor or crane load to gain access to the roof.

- Lower material from the roof deck to the ground in a safe manner to avoid injuries or property damage.

- Establish approved "drop zones" where walking or working is prohibited. An approved drop zone cannot be located above an entrance or exit to a home.

- Provide a spotter on the ground level to create an added layer of protection to prevent anyone from walking into the drop zone area.

Roof sheathing operations

- Do not install roof sheathing unless you are qualified to do so.

- Employees who are not involved in the roof sheathing installation should not stand or walk below or adjacent to the roof opening or exterior walls or in any area where they could be struck by falling objects.

- The competent person must define the limits of this area before sheathing begins.

- The competent person must stop work as needed to allow passage through such areas when this work stoppage would not create a greater hazard.

- The bottom row of roof sheathing may be installed by workers standing in truss webs.

When using slide guards:

- After the bottom row of roof sheathing is installed, a slide guard extending the width of the roof must be securely attached to the roof.

- Slide guards must be constructed of 2 × 4 or 2 × 6 lumber capable of limiting an uncontrolled slide.

- Install the slide guard while standing in truss webs and leaning over the sheathing.

- Additional rows of roof sheathing may be installed when you are positioned on previously installed rows of sheathing. A slide guard can help you retain your footing during successive sheathing operations.

- Additional slide guards must be securely attached to the roof at intervals not to exceed 13 ft. as successive rows of sheathing are installed.

- Roofs with pitches in excess of 9:12 must have slide guards installed at 4 ft. intervals.

- In wet weather (rain, snow, or sleet), roof sheathing operations must be suspended unless safe footing can be ensured.

- Suspend roof-sheathing operations during winds above 40 mph unless windbreakers are erected.

Roof truss/rafter erection

- Employees must be trained in the safe work practices for working on or in the truss.

- Employees should not have other duties while setting trusses or erecting rafters.

- Once truss or rafter installation begins, employees who are not involved in that activity must not stand or walk where they could be struck by falling objects.

- The first two trusses or rafters should be installed from scaffolding or ladders and secured in place.

- Ensure that the wall can support the weight of the ladder leaning on side walls or a scaffold that may be attached to a wall.

- After the first two trusses or rafters have been set, an employee can climb onto the interior wall top plate or previously stabilized trusses to brace or secure the peaks.

- Safe access to the truss/rafter must be provided.

- Employees should use the previously stabilized truss/rafter as a support, while setting each additional truss/rafter.

- All trusses/rafters must be adequately braced before any employee can use the truss/rafter as a support.

- Employees can only move onto the next truss/rafter after it has been secured.

- Walking and working on the exterior wall top plate is not permitted.

- Employees positioned at the peak or in the webs of the trusses must work from a stable position either by sitting on a ridge seat or by positioning themselves in a previously stabilized truss or rafter.

- Employees should not stay on the peak/ridge any longer than necessary to safely complete the task.

Follow these additional safe work practices when working inside the trusses.

- Only trained employees may be allowed to work at the peak during roof truss or rafter installation.

- Employees may perform work inside truss areas such as installing bracing or making repairs.
- All trusses should be adequately braced before any employee can use the truss as a support.
- Safe access into the truss area must be provided.
- When practical, temporary flooring should be installed to create a safe, solid work platform if ongoing work will be performed in the attic area.
- Keep both hands free when climbing through the truss areas.
- In order to avoid laceration injuries, do not place hands near metal gusset plates of roof trusses.

Employees who are securing and bracing trusses or detaching trusses from a crane should

- have no other duties
- work at the peaks or in the webs of trusses in a stable position by working in previously braced trusses
- not remain on the peak or in the truss any longer than necessary to safely complete the task

Interior scaffolding, such as sawhorse or trestle ladder, or exterior scaffolding, such as top plate, can be used as a safe alternative when setting trusses or erecting rafters. Various types and brands of interior and exterior scaffolding are commercially available. Be sure to follow the manufacturer's safety instructions for all scaffolding. Interior scaffolding can be installed along the interior wall below the area where the trusses/rafters will be set and work can be done from this established platform. For exterior scaffolding, walls that support the scaffold must be capable of supporting, without failure, their own weight plus four times the maximum intended load on the scaffolding.

Key elements of scaffolding safety include the following:

- Scaffolding that is 10 ft. or higher must be equipped with guardrails.
- Scaffolding planks must be secured so that they do not move.
- Interior scaffolding may consist of sawhorse scaffolding, planks supported by ladders, or planks supported by top plate brackets.
- Scaffolding set up and take down must be conducted under the supervision of a competent person.

Another option for safe installation of trusses is to position the truss system on the ground. Sheath and brace the trusses for stability. The fully assembled roof system can be lifted and set in place with a forklift or crane. Remember to follow the manufacturer's instructions for safe forklift and crane operation.

Include the following safe work practices for materials and equipment specific to the jobsite. You may need to modify these safe work practices to meet your company's specific needs.

SAFE WORK PRACTICES FOR CRANES

All employees must observe the following safe work practices.

- Do not walk under crane loads.
- Follow the crane operator's and signaler's instructions.
- Stay clear of the crane's superstructure.
- Stay a safe distance away from cranes in operation if your work does not require you to be in the area.
- Hand signals used for crane operations must meet the ANSI standard for crane type used and an illustration of the signals must be posted at the jobsite.

Include the following safe work practices for materials and equipment specific to the jobsite. You may need to modify these safe work practices to meet your company's specific needs.

SAFE WORK PRACTICES FOR MOTOR VEHICLES

All employees must observe the following safe work practices.

- Ensure that all off-road equipment used on site is equipped with rollover protection.
- Use a backup alarm or guide when operating in bidirectional machines such as rollers, compactors, front-end loaders, bulldozers, and similar equipment.
- Use a horn that can be distinguished from the surrounding noise when operating motor vehicles.
- Maintain the horn in operating condition.
- Ensure that all vehicles have fully operational braking systems and brake lights.
- Ensure that all vehicles have working headlights and taillights.
- Use seat belts when transporting employees in motor and construction vehicles.
- Seats must be firmly secured with enough seating for all employees being transported.
- Inspect all vehicle brake systems, tires, horns, steering mechanisms, lights, couplings, and seat belts (not required for equipment used for stand-up operation) before use for broken or unsafe conditions.

Include the following safe work practices for materials and equipment specific to the jobsite. You may need to modify these safe work practices to meet your company's specific needs.

SAFE WORK PRACTICES FOR EXCAVATION AND TRENCHING

All employees must observe the following safe work practices.

Protective systems

- Protect all excavations over 5 ft. deep with shoring, trench boxes, or sloping of the earth according to the direction of the competent person.
- Protect excavations less than 5 ft. deep with shoring, trench boxes, or the sloping of the earth if the competent person determines that a cave-in is a possibility.

General guidelines

- Remove or support all surface obstructions as needed.
- Request utility companies or owners to determine the location of underground utility installations before excavating.
- Ensure that underground installations are protected, supported, or removed while the excavation is open.
- Proceed with caution and use detection equipment if the exact location of lines is unknown.
- Use stairs, ladders, and ramps every 25 ft. to exit the excavation.
- Wear warning vests or other high-visibility materials when near traffic.
- Do not stand beneath loads handled by lifting or digging equipment.
- Keep clear of vehicles loading or unloading.
- Use a warning system (barricades, hand or mechanical signals, or stop logs) if the equipment operator cannot see near the edge of an excavation.
- Do not work in excavations where water is standing or accumulating unless special supports are used to prevent cave-ins.
- Keep all spoils piles and equipment at least 2 ft. away from the edge of the excavation.
- Use support systems such as shoring, bracing, or underpinning to protect all other structures from collapse.
- Remove loose soil or use barricades for protection inside excavations.
- Inspect excavations, adjacent areas, and protective systems for
  - evidence of possible cave-ins
  - indications of failure of protective systems
  - hazardous atmospheres
  - other hazardous conditions at the beginning of each workday or if the stability of the excavation could change
- Install walkways over excavations that must be crossed. A guardrail system must be in place if the walkway is 6 ft. or more above the lower level.
- Barricade wells, pits, or shafts.

Include the following safe work practices for materials and equipment specific to the jobsite. You may need to modify these safe work practices to meet your company's specific needs.

SAFE WORK PRACTICES FOR STAIRWAYS AND LADDERS

All employees must observe the following safe work practices.

General requirements for temporary stairs

- Use a stairway, ramp, or ladder whenever you must step more than 19 in. to gain access to another surface.

- Build temporary stairs between 30 and 50 degrees from horizontal.

- Make riser height and tread depth the same for each flight of stairs. Don't vary by more than ¼ in.

- Provide a platform when doors open directly on a stairway, and make sure the width of the platform is not reduced to less than 20 in.

- Keep hazardous projections, such as protruding nails, splinters, etc., out of the stairs.

- Fix slippery conditions before the stairs are used.

- Place wood or other materials in unfilled pan stairs to keep the surface even.

- Build treads of stairs with wood or other solid material and install them the full width and depth of the stair.

Stair rails and handrails

- Install a 36-in.-high stair/handrail on unprotected sides of stairs with more than 3 risers or more than 30 in. high. Measure the 36 in. from the tread surface up to the top of the rail.

- Install midrails halfway between the top edge of the stair rail system and the stairway steps.

- Use a handrail offset on winding/spiral stairs to keep employees from walking where the treads are less than 6 in. wide.

- Build stair/handrail systems to withstand 200 lb. of force in any downward or outward direction at any point along the top edge.

- Install handrails to act as a handhold between 30 and 37 in. on all stair systems even if there is a wall.

- Surface stair/handrails to prevent injury and clothes snagging.

- Build stair handrails in a manner that does not create a projection hazard beyond the edge of the rail.

- Build temporary stair/handrails about 3 in. away from where the drywall will be installed, so that the rails will not have to be taken down in order to install the drywall.

- Use guardrail system to protect sides and edges of stairway landings.

Ladders

The following safety requirements apply to all ladders, including job-made ladders:

- Use ladders only on stable and level surfaces unless they are secured to prevent movement.

- Keep the area around the top and bottom of ladders clear of debris.
- Place ladders on nonslip surfaces or use ladders that have slip-resistant feet.
- Inspect ladders before each use and after any activity that could have caused damage.
- Remove ladders with structural defects from service.
- Maintain ladders free of oil, grease, and other slipping hazards.
- Follow the manufacturer's rated capacity for ladder use.
- Use ladders only for their designated purpose.
- Do not move, shift, or extend ladders that are occupied.
- Do not work outside of the foot print of the ladder. Never let your belt buckle pass beyond either ladder siderail.
- Place ladders at an angle of 4:1 for every 4 ft. in height the ladder should be placed 1 ft. out.
- Extend ladders at least 3 ft. beyond the surface you are accessing so that you have a handhold when getting on and off.
- Do not use a stepladder as an extension ladder.
- Do not tie ladders together to create a longer ladder.
- Do not put ladders on scaffolds or other unstable platforms.
- Support both ladder rails equally.
- Keep ladders from coming within 10 ft. of energized power lines.
- Do not use the top two steps of a step ladder as a seat or a ladder.
- Do not climb the cross bracing on the back of stepladders.
- Use the three-point contact system when working on ladders. Face the ladder when climbing up or down, and use at least one hand to hold the ladder.
- Do not carry objects or loads that could cause you to fall.
- Be sure that ladder rungs and steps are parallel, level, and uniformly spaced.
- Be sure that rungs or steps of portable ladders or stepladders are between 10 and 14 in. apart.
- Be sure that side rails for all portable ladders are at least 12 in. apart.
- Always use and lock the metal spreader or device that holds the front and back sections apart.
- Surface ladder components to prevent punctures, lacerations, or clothes snagging.
- Coat wood ladders with clear covering only. Identification or warning labels can be placed on one face of a side rail.
- Do not build or use job-made ladders unless they meet the exact ANSI and OSHA specifications.

Include the following safe work practices for hazard communication practices specific to the jobsite. You may need to modify these safe work practices to meet your company's specific needs.

SAFE WORK PRACTICES FOR HAZARDOUS COMMUNICATION

This company has

- obtained a copy of the hazard communication standard
- read and understood the requirements
- assigned responsibility for tasks (maintaining MSDSs, maintaining labels, and conducting training)
- prepared an inventory of chemicals
- ensured containers are labeled
- obtained MSDS for each chemical
- prepared a written program
- made MSDSs available to workers
- conducted employee training
- established procedures to maintain current program
- established procedures to evaluate the effectiveness of our written program

COMPANY SAFETY EDUCATION AND TRAINING

Supervisory training

Safety education of all employees, from superintendents to field employees, will be conducted through all phases of the work performed by the company.

Employee safety rules

Employees will receive training on the company's job rules and regulations and the employee's personal safety requirements. Each jobsite employee will be required to sign and date an acknowledgement of receipt of the safety program and policies.

Safety toolbox talks

Weekly toolbox talks will be conducted and last approximately 15 minutes. The talks will include time for active participation by employees, including a question-and-answer session. Toolbox talks will also be scheduled at the beginning of new operations to ensure that employees are familiar with safe work practices and the requirements of upcoming work.

FIELD CONSTRUCTION SAFETY AND HEALTH REQUIREMENTS

Jobsite safety inspections

Inspections will be conducted periodically to identify and correct unsafe practices and conditions. These inspections will focus on the identification and correction of potential safety, health, and fire hazards. As part of an effective inspection program, the site superintendent will

- set inspection responsibilities and schedules
- develop an administrative system for review of reports
- set up a procedure to follow up on corrected conditions
- analyze inspection findings
- set program standards for observing employee safety practices
- communicate program standards for observing employee practices to each supervisor

- communicate program safety standards to workers
- monitor performances of workers' safety practices

Weekly jobsite safety audit

The site superintendent will perform a weekly jobsite safety inspection using the safety audit checklist. After completing and signing the form, the site superintendent will submit it to the president/owner. The site superintendent will conduct the audit in person and will not delegate the audit to other employees.

President/owner inspections

The president/owner and the site superintendent will discuss the status of site safety and loss-control programs and performance results to date, as measured against the company's targeted goals on every jobsite visit. The president/owner will tour all company work locations with the site superintendent to review jobsite working conditions and compliance with company safety policies.

Company insurance carrier safety audits

Carriers of the company's workers' compensation, general liability, and automobile insurance may need to conduct a jobsite safety inspection or accident investigation. These company insurance safety audits must be scheduled with the approval of the president/owner, who will notify the site superintendent when the insurance representatives will be on site. Site supervisory personnel are expected to fully cooperate with the company's insurance representatives.

EMERGENCY PREPAREDNESS PROCEDURES

Medical and first aid requirements

Telephone numbers for emergency service units will be posted on the jobsite. In the event of any emergency, the site superintendent will render first aid and CPR, if qualified, until medical emergency personnel arrive. OSHA requires that at least one person be trained to render first aid on a jobsite in the absence of a nearby treatment facility.

Fire prevention and protection

The site superintendent is knowledgeable of and in compliance with fire prevention and protection regulations. Fire protection refers to the use of fire extinguishers, evacuation routes, and emergency procedures when a fire does occur. Company new hire training sessions will include:

- emergency telephone numbers
- locations of fire alarm systems throughout the jobsite
- location and proper operation of fire extinguishers
- emergency evacuation routes and procedures
- procedures to account for all employees once evacuated

Model Written Hazard Communication Program

This model hazard communication program should be used only as a guide. Because individual company circumstances vary widely, you should modify the program to reflect actual company operations. This program does not contain legal advice. You should consult with your attorneys or insurance company representatives before acting on the premises described herein.

OUR HAZARD COMMUNICATION PROGRAM

(<u>Name of company</u>) is firmly committed to providing all of its employees with a safe and healthy work environment. It is a matter of company policy to provide our employees with information about hazardous chemicals on the work site through our hazard communication program that includes container labeling, material safety data sheets (MSDSs), and employee information and training.

(<u>Name of person or position</u>) will have the overall responsibility for coordinating the hazard communication program for (<u>Name of company</u>).

(<u>Name of person or position</u>) will make our written hazard communication program available, upon request, to employees, their designated representatives, the Assistant Secretary of Labor for Occupational Safety and Health, and the Director of the National Institute for Occupational Safety and Health.

LIST OF HAZARDOUS CHEMICALS

(<u>Name of person or position</u>) will compile a list of all hazardous chemicals that will be used on the work site by reviewing container labels and MSDSs. The list will be updated as necessary. It will be kept (<u>location</u>). *[Be sure to attach your list of hazardous chemicals.]*

LABELING

It is our policy to ensure that each container of hazardous chemicals that is present on a jobsite is properly labeled. The labels will list:

- the contents of the container
- appropriate hazard warnings
- the name and address of the manufacturer, importer, or other responsible party

To further ensure that employees are aware of the chemical hazards of materials used in their work areas, it is our policy to label all secondary containers. Secondary containers will be labeled with either an extra copy of the manufacturer's label, or with a sign or generic label that lists the container's contents and appropriate hazard warnings.

This responsibility has been assigned to (name of person).

MATERIAL SAFETY DATA SHEETS (MSDS)

Copies of MSDSs for all hazardous chemicals to which employees may be exposed are kept (location) and are readily accessible to employees in the work area during each work shift. (Name of person or position) is responsible for obtaining, maintaining, and updating the file of MSDSs.

EMPLOYEE TRAINING

Employees are to attend a training session on hazardous chemicals in their work area at the time of their initial work assignment. The training session will cover the following:

- an overview of the hazard communication requirements
- chemicals present on our jobsites
- the location, availability, and contents of our written hazard communication program and MSDSs
- methods to detect the release or presence of hazardous chemicals in the work area
- physical and health hazards of the chemicals in the work area
- methods to lessen or prevent exposure to hazardous chemicals by using good work practices, personal protective equipment, etc.
- emergency first aid procedures

When a new type of product is introduced into a work area or the chemical composition of a product changes, (person or position) will review the above items as they are related to the new chemicals.

NONROUTINE TASKS

Periodically, employees are required to perform non-routine tasks. Prior to starting work on such projects, each affected employee will be informed by (person or position) about hazards to which they may be exposed and appropriate protective and safety measures.

INFORMING OTHER EMPLOYERS

To ensure that the employees of other contractors have access to information on the hazardous chemicals at a jobsite, it is the responsibility of (person or position) to provide the other contractors with the following information:

- location of MSDSs
- the name and location of the hazardous chemicals to which their employees may be exposed
- appropriate protective measures required to minimize exposure
- an explanation of the labeling system used at the jobsite

Each contractor bringing chemicals onto a jobsite must provide this company with the appropriate hazard information on those substances to which our own employees may be exposed on a jobsite.

Model Hazard Communication Employee Training Program

This sample hazard communication employee training program should be used only as a guide. Because individual company circumstances vary widely, you should modify the program to reflect actual company operations. This program does not contain legal advice. You should consult with your attorneys or insurance company representatives before acting on the premises described herein.

OUR EMPLOYEE HAZARD COMMUNICATION TRAINING PROGRAM

Our Hazard Communication Standard (HazCom) is intended to ensure that both employers and employees are aware of potential hazards associated with chemicals in the workplace. We use a variety of products on our jobsites. Many of these products contain one or more hazardous chemicals. Most of the products can be grouped by their basic function or use. Our training teaches employees which products fit in each group, how to identify the associated hazards, and how to detect and control these hazards. A list of the chemicals potentially found on our jobsites is attached to the company's written HazCom program.

The training covers the following:

- chemicals present on our jobsites
- the location, availability, and contents of our written hazard communication program and MSDSs
- methods to detect the release or presence of hazardous chemicals in the work area
- physical and health hazards of the chemicals in the work area
- methods to lessen or prevent exposure to hazardous chemicals by using good work practices, personal protective equipment, etc.
- general emergency and first aid procedures
- company safe work practices
- methods to detect the presence of a chemical

MATERIAL SAFETY DATA SHEETS (MSDSs) AND LABELS

Although MSDSs can appear in many different formats, the information they contain will essentially be the same. Employees should use the MSDS as reference information for the product label. An MSDS contains the following fields of information:

Substance Identification

- chemical name (as it appears on the label)

- manufacturer's name and address
- emergency telephone numbers
- date the MSDS was prepared and the signature of the preparer

Hazardous Ingredients/Identity Information

- **Hazardous components.** Contains the specific chemical identity, its formula, and any common name by which it is known.

- **OSHA Permissible Exposure Limits (PEL).** Indicates the permissible maximum amount of the chemical a person may be safely exposed to without harm.

- **American Conference of Governmental Industrial Hygienists Threshold Limit Value (TLV).** Indicates the concentration of a chemical in the air that can be breathed for five consecutive eight-hour workdays by most people without harmful effects. It is generally expressed in parts per million.

- **Other limits recommended.** Indicates the recommended limits on the use of the chemical by any agency, scientific group, or organization.

Physical/Chemical Characteristics

- *Boiling point.* The temperature at which a liquid boils.

- *Vapor pressure (mm Hg).* Measures a liquid's tendency to evaporate. The higher the pressure, the faster it will evaporate.

- *Vapor density.* Indicates the weight of the vapor compared to the weight of an equal volume of air. If a vapor is heavier than air (vapor density greater than 1), then it will sink to the ground. If it is lighter than air (vapor density less than 1), then it will rise. For example, with flammable materials, when the vapor density is greater than 1, vapors tend to collect in the lowest spot. A contractor must be alert to vapors traveling to an ignition source and then flashing back to the vapor source. Care must also be taken to ensure that vapors do not displace oxygen.

- *Solubility in water.* Indicates whether chemical can mix with water in any ratio without separating.

- *Appearance and odor.* Is a brief description of the chemical's color and smell.

- *Specific gravity.* Is the ratio of the weight of the material to the weight of an equal volume of water. The specific gravity values less than or equal to 1 indicate that water should not be used to extinguish a fire involving the substance unless the water comes from automatic sprinklers.

- *Melting point.* Indicates the temperature at which a solid changes to a liquid.

- *Evaporation rate (butyl acetate = 1).* Indicates the temperature at which a substance evaporates.

Fire and Explosion Hazard Data

- *Flash point.* Indicates the lowest temperature at which a liquid gives off enough vapor to ignite in air when exposed to a flame. When the flash point is between 100° and 110° F, extra care must be taken in hot environments because the temperature of the vapor could be high enough to ignite if an ignition source is introduced. Such sources might be cigarette smoking, electrical equipment and wiring, cutting and welding, or static electricity. A red diamond is required on all liquids that OSHA has classified as flammable (flash point values of 99.9° F or below).

- **Flammable limits.** Indicates the range of vapor concentrations that will explode when an ignition source is present. The lower explosive limit (LEL) is the minimum amount of vapor in the air that can be ignited. The upper explosive limit (UEL) is the maximum amount of vapor in the air that will sustain fire.

- **Extinguishing media.** Indicates the materials suitable for extinguishing a fire involving the identified chemical. These firefighting agents are water, fog, foam, alcohol foam, carbon dioxide, and dry chemical.

- **Special firefighting procedures.** Indicates the chemical's special characteristics when it comes in contact with fire, such as whether it is difficult to extinguish, will reignite spontaneously, or can be extinguished by water or other firefighting agents. This subsection will also indicate any required protective equipment needed when fighting the fire and evaluate any toxicity of the material on anyone fighting the fire.

- **Unusual fire and explosion hazards.** Indicates any special types of hazards requiring attention, such as whether the chemical is difficult to extinguish or will reignite spontaneously and how it reacts with water and other extinguishing agents. For example, if water is applied to a combustible liquid with a flash point above 212° F, it may foam violently or boil over, endangering workers and firefighters.

Reactivity Data

- **Stability.** Indicates conditions that contribute to the stability or instability of a chemical when it is exposed to heat, pressure, or excessive shock during storage, use, misuse, or transport. Refer to this section to identify specific conditions to be avoided. The warnings, for example, may read, "reacts violently with water" or "avoid sudden shock."

- **Incompatibility (materials to avoid).** Indicates various materials or conditions you must keep the chemical away from to avoid adverse reactions, such as a substance that ignites or explodes when it comes in contact with the chemical.

- **Hazardous decomposition or by-products.** Indicates gases or vapors that are released when the chemical burns or decomposes. It details the hazardous substances your employees may be exposed to as a result of heating, handling, or burning the chemical.

- **Hazardous polymerization.** Occurs when molecules of the chemical combine with molecules of another material to form a larger, different material. This reaction is accompanied by the release of large amounts of energy that can produce fire or other hazards. Polymerization can occur when the chemical comes in contact with certain plastics, rubber, or coatings. This section of the MSDS will indicate possible storage conditions that could result in polymerization. It will also indicate any inhibitors—chemicals that can be added to prevent or delay polymerization—and the expected time period in which an inhibitor is expended.

Health Hazard Data

- **Route(s) of entry.** Indicates how a chemical may enter the body either through inhalation, contact with the skin or eyes, or ingestion.

- **Health hazards.** Indicates any long-term (chronic) or short-term (acute) effects of a chemical on the human body.

- **Carcinogenicity.** Indicates whether the chemical causes cancer. It is important that your employees understand that not all hazardous substances cause cancer.

- **Signs and symptoms of exposure.** Describe the effects of exposure to the chemical, such as an employee's appearance and the most common resulting sensations, for example, headache, dizziness, or nausea.

- **Medical conditions severely aggravated by exposure.** Indicates how the chemical will affect any preexisting medical conditions.

- **Emergency and first aid procedures.** Indicates how to reduce the hazardous chemical effects when an employee has been exposed through inhalation or skin or eye contact. These are emergency procedures only. Exposed employees should be examined by a doctor as soon as possible.

Precautions for Safe Handling and Use

- **Steps to be taken in case material is released or spilled.** Indicate precautions, such as avoiding breathing gases and vapors, avoiding contact with liquids and solids, removing ignition source, and use of special equipment for cleanups. This section also provides techniques for controlling land or water spills.

- **Waste disposal methods.** Details proper disposal of the chemical and contaminated materials.

- **Handling and storage precautions.** Emphasizes incompatibility or polymerization problems that could occur during storage or handling of the chemical.

- **Other precautions.** Indicates other special precautions for handling or disposal of the chemical.

Control Measures

- **Respiratory protection.** Specifies type of respirators required by OSHA when the chemical is used even as a precautionary measure in non-emergency situations.

- **Ventilation.** Indicates necessary ventilating systems to prevent overexposure to the chemical. The objective is to prevent the substance from reaching the employee's breathing zone. A local exhaust ventilation system has a high speed and low volume that can capture a chemical quickly after it has been released. A mechanical (or general) ventilation system can be used to heat and/or cool an enclosed area in a permanent facility.

- **Protective gloves.** Indicates whether gloves must be worn when the chemical is being handled. If gloves are required for skin protection, the type of material that the gloves should be made of will be indicated.

- **Eye protection.** Indicates appropriate eye protection, such as face shields, safety goggles, or glasses.

- **Other protective clothing or equipment.** Indicates protective equipment, such as aprons or boots, and the materials the equipment should be made of to effectively prevent skin contact.

Labels

Employees should read product labels before working with a hazardous substance. Each label will include:

- the contents of the container
- appropriate hazard warnings
- the name and address of the manufacturer, importer, or other responsible party

To further ensure that employees are aware of the chemical hazards of materials used in their work areas, all secondary containers will also be labeled with either an extra copy of the manufacturer's label, a sign, or generic label that lists the contents and appropriate hazard warnings. The label should serve as a reminder of the information we are presenting in this training session and of the information found in more detail on the MSDS. It is essential that employees read the hazard warning and use the chemical as prescribed by the label.

About the National Association of Home Builders

The National Association of Home Builders is a Washington-based trade association representing more than 235,000 members involved in home building, remodeling, multifamily construction, property management, trade contracting, design, housing finance, building product manufacturing, and other aspects of residential and light commercial construction. Known as "the voice of the housing industry," NAHB is affiliated with more than 800 state and local home builders associations around the country. NAHB's builder members construct about 80 percent of all new residential units, supporting one of the largest engines of economic growth in the country: housing.

Join the National Association of Home Builders by joining your local home builders association. Visit www.nahb.org/join or call 800-368-5242, x0, for information on state and local associations near you. Great member benefits include:

- Access to the **National Housing Resource Center** and its collection of electronic databases, books, journals, videos, and CDs. Call 800-368-5254, x8296 or e-mail nhrc@nahb.org
- **Nation's Building News**, the weekly e-newsletter containing industry news. Visit www.nahb.org/nbn
- **Extended access to www.nahb.org** when members log in. Visit www.nahb.org/login
- **Business Management Tools** for members only that are designed to help you improve strategic planning, time management, information technology, customer service, and other ways to increase profits through effective business management. Visit www.nahb.org/biztools
- **Council membership:**
  - **Building Systems Council:** www.nahb.org/buildingsystems
  - **Commercial Builders Council:** www.nahb.org/commercial
  - **Building Systems Council's Concrete Home Building Council:** www.nahb.org/concrete
  - **Multifamily Council:** www.nahb.org/multifamily
  - **National Sales & Marketing Council:** www.nahb.org/nsmc
  - **Remodelors™ Council:** www.nahb.org/remodelors
  - **Women's Council:** www.nahb.org/womens
  - **50+ Housing Council:** www.nahb.org/50plus

BuilderBooks, the book publishing arm of NAHB, publishes inspirational and educational products for the housing industry and offers a variety of books, software, brochures, and more in English and Spanish. Visit www.BuilderBooks.com or call 800-223-2665. NAHB members save at least 10% on every book.

BuilderBooks Digital Delivery offers over 30 publications, forms, contracts, and checklists that are instantly delivered in electronic format to your desktop. Visit www.BuilderBooks.com and click on Digital Delivery.

The **Member Advantage Program** offers NAHB members discounts on products and services such as computers, automobiles, payroll services, and much more. Keep more of your hard-earned revenue by cashing in on the savings today. Visit www.nahb.org/ma for a comprehensive overview of all available programs.